NELSON
MASTERING Math
POST-SECONDARY PREPARATION

Jack Weiner
Professor of Mathematics Emeritus, University of Guelph

Gord Clement
Upper Grand District School Board

COPYRIGHT © 2017 by
Nelson Education Ltd.

ISBN-13: 978-0-17-682347-4
ISBN-10: 0-17-682347-6

Printed and bound in Canada
1 2 3 4   20 19 18 17

ALL RIGHTS RESERVED. No part of this publication may be reproduced, stored in a retrieval system, or transmitted in any form or by any means, electronic, mechanic, photocopying, scanning, recording or otherwise, except as specifically authorized.

Every effort has been made to trace ownership of all copyrighted material and to secure permission from copyright holders. In the event of any question arising as to the use of any material, we will be pleased to make the necessary corrections in future printings.

Cover Design
Trinh Truong

Cover Image
Rubberball Productions/Brand X Pictures/Getty Images

TABLE OF CONTENTS

How to Get an "A" in Math! 1
How to Get Extra Help 2

Operation Cooperation and Fraction Traction

BEDMAS (Order of Operations) 3
Adding and Subtracting Fractions 4
Multiplying and Dividing Fractions 5
Operations with Decimals 6
Ratio, Rate, and Percent 9
Complex Numbers 10

Factoring: A Product of Practice

Factoring Difference of Squares 11
Factoring Difference of Cubes 12
Factoring $a^n - b^n$ and $a^n + b^n$ 15
Common Factors 16
Factoring Easy Trinomials 17
The Remainder and Factor Theorems for Polynomials 18

Pliable Polynomials

Multiplying Expressions—FOIL 21
Adding and Subtracting Polynomial Fractions 22
Multiplying and Dividing Polynomial Fractions 23
Polynomial Division 24

A Partial Look into Partial Fractions

Partial Fractions: Preliminaries 27
Partial Fractions: Distinct Linear Factors 28
Partial Fractions: Repeated Linear Factors 29
Partial Fractions: Irreducible Quadratics 30

The Straight Goods on Lines

Finding the Equation of a Line 33
Slope m and y Intercept b 34
Graphing a Straight Line Using $y = mx + b$ 35
Distance between Two Points and Distance from a Point to a Line or Plane 36
Visually Identifying Slopes of Lines 39
Parallel and Perpendicular Lines 40
Finding Tangent and Normal Lines to a Curve 41

A Few Lines on Linear Algebra

Solving a Linear Equation	42
Solving Two Linear Equations Using Substitution	45
Solving Two Linear Equations Using Row Reduction	46
Solving Three Linear Equations Using Row Reduction	47
Solving Three Linear Equations REDUX: MATRICES	48
Consistent, Inconsistent, and Dependent Systems of Linear Equations	51

Giving the Third Degree to Second Degree Polynomials: Quadratics!

Solving Quadratic Equations Using the Quadratic Formula	52
Factoring Quadratic Expressions Using the Quadratic Formula	53
Problems Involving the Sum and Product of the Roots of a Quadratic Equation	54
The Graph of $y = a(x-b)^2 + c$	57
Completing the Square	58

Solving Inequalities with Less (<) Difficulty, Greater (>) Ease

Solving Linear Inequalities	59
Solving Quadratic Inequalities	60
Solving Inequalities with Two or More Factors	63
Solving Rational Inequalities	64

Increasing the Magnitude of Your Absolute Value Knowledge

The Basics of Absolute Value	65
Solving Absolute Value Equations	66
Solving Easy Absolute Value Inequalities	69
Solving Less Easy Absolute Value Inequalities	70

Getting to the Root of Square Roots

The Basics of $\sqrt{}$ and the Reason $\sqrt{x^2} = |x|$	71
Solving Equations Involving Square Roots	72
Rationalizing Denominators that Have $\sqrt{}$	75

Some Basic Graphs and Some Basics about Graphs

Graphs of Basic Quadratic Relations	76
Basic $y = x^n$ Graphs, where $n \in N$ (Even and Odd Functions)	77
Basic $y = \dfrac{1}{x^n} = x^{-n}$ Graphs, where $n \in N$	78
Basic $y = x^{\frac{1}{n}}$ Graphs, where $n \in N$	81
Shifting or Rescaling a Given Graph	82

Tests for Symmetry ... 83
Graphing Polynomials Without Calculus ... 84
Vertical and Horizontal Asymptotes ... 87
Slant Asymptotes ... 88
Intersection of Two Curves ... 89
The Greatest Integer (or Floor) Function ... 90
Graphs with the Greatest Integer Function ... 93

The Mastering Math Logs Powerful Time with Exponents and Logarithms

Properties of Exponents ... 94
Logarithms (Log Means "FIND THE EXPONENT!") ... 95
Basic Exponential Graphs ... 96
Basic Logarithm Graphs ... 99
Inverse Formulas for Exponents and Logarithms ... 100
Solving Exponential Equations ... 101
Solving Logarithm Equations ... 102
The Derivative of $y = e^x$ and $y = a^x$ ... 105
The Derivative of $y = \ln x$ and $y = \log_a x$ ... 106
Logarithmic Differentiation Part I ... 107
Logarithmic Differentiation Part II: $\frac{d}{dx}\left(f(x)^{g(x)}\right)$ ... 108
Integrals Yielding ln: $\int \frac{\frac{du}{dx}}{u} dx = \ln |u| + C$ ... 111

Drawing Your Attention to Some Basic Geometry

A Degree of Knowledge about Angles ... 112
The Pythagorean Theorem ... 113
Similar Triangles ... 114
Radian Measure of an Angle ... 117

Angling Right in on Trigonometry

Basic Trigonometric Ratios: SOH CAH TOA ... 118
Using SOH CAH TOA to Find Missing Sides and Angles ... 119
Angles in Standard Position ... 120
Related Angles in Standard Position ... 123
Trig Ratios for the $(30°, 60°, 90°) \equiv (\pi/6, \pi/3, \pi/2)$ Triangle ... 124
Trig Ratios for the $(45°, 45°, 90°) \equiv (\pi/4, \pi/4, \pi/2)$ Triangle ... 125
Trig Ratios for 30°, 45°, 60° (and More)—A Table ... 126
Trig Ratios for 30°, 45°, 60° (and More)—A (Fabulous) Picture ... 129
Basic Trigonometric Graphs ... 130

The Circle Definition of Sine and Cosine 131
Solving the Trig Equation sin $x = c$ 132
Solving the Trig Equation cos $x = c$ 135
The Sine Law 136
The Cosine Law 137
Commonly Used Trigonometric Formulas Including Derivatives and Integrals 138
Basic Inverse Trigonometric Graphs 141

A Straightforward Approach to Limits

Easy Limits: "No Problem" Problems 142
"0/0" Limits 143
One-Sided Limits 144
Limits which Approach $\pm\infty$ 147
Limits at Infinity 148
An "$\infty - \infty$" Limit: $\lim\limits_{x \to \infty}\left(\sqrt{x^2 + 8x} - x\right)$ 149
Limits: A Summary 150
Variations on $\lim\limits_{\theta \to 0}\dfrac{\sin \theta}{\theta} = 1$ 153
L'Hôpital's Rule 154
L'Hôpital's Rule Disguised: Converting IFs to Fractions 155

Continuity

Domain (Food for a Function!) 156
Composite Functions 159
Continuity and Discontinuity at a Point 160
Continuous Functions (Intervals of Continuity) 161
Continuity and Branch Functions 162
Essential versus Removable Discontinuities 165

Derivatives or Going on a Tangent about Slopes

Finding the Derivative from the Definition 166
Differentiable Functions (Intervals of Differentiability) 167
Differentiability and Branch Functions 168
Critical Numbers 171
Max and Min Points from the First Derivative 172
Graphing y vs y' vs y'' (Increasing/Decreasing and Concavity) 173
Graph Sketching with Calculus 174
Graph Sketching with Calculus: Vertical Tangent! 177
Estimating Using the Differential 178
Rolle's Theorem 179
The Mean Value Theorem 180

Derivative Rules Rule

Derivatives: The Product Rule 183
Derivatives: The Chain Rule 184
Derivatives: The Quotient Rule 185
Derivatives: Implicit Differentiation 186
Derivatives: Implicit Differentiation Second Derivative 189

Integrating Your Knowledge about the Anti-Derivative

Easy Integrals/Anti-Derivatives 190
Easy Integrals that Need a Little Tweaking 191
The **C**hain **R**ule **I**n **R**everse: No Adjustments Needed 192
CRIR: Adjustments Needed BUT Don't Use Substitution! 195
CRIR: Adjustments Needed and Using Substitution 196
Substitution when the CRIR Doesn't Apply 197
CRIR: Products of Trig Functions 198
Integration by Parts: The Basic Examples 201
Integration by Parts: Circular Integration By Parts 202
Integration by Parts (I by P): The Tan—Sec Connection 203
Integration by Trigonometric Substitution: Sin 204
Integration by Trigonometric Substitution: Tan 207
Integration by Trigonometric Substitution: Sec 208
Integration Using Partial Fractions 209
Definite Integrals – Area Problems 210
Definite Integrals Using Substitution 213
Improper Integrals – Functions with a Discontinuity 214
Improper Integrals – Infinite Limits of Integration 215
The Derivative of an Integral 216
Differential Equations – Separation of Variables 219

Inverse Functions: Now That's a Switch!

Finding the Inverse of a Function 220
Derivatives of Inverse Functions 221

Parametric Equations: Making Relations Functional

Parametric Equations 222
Derivatives from Parametric Equations 225
Higher Derivatives from Parametric Equations 226

Warming Up to Polar Coordinates

Polar Coordinates 227
Polar to Rectangular Coordinates; Rectangular to Polar Equations 228
Rectangular to Polar Coordinates; Polar to Rectangular Equations 231

Going to Any Lengths to Give You New Direction with Vectors

(Very) Basic Vectors	232
The Dot or Scalar or Inner Product of Two Vectors	233
The Projection of One Vector on Another	234
The Vector or Cross Product of Two Vectors	237
The Vector Equation of a Line	238
The Vector Equation of a Plane	239
The Scalar Equation of a Plane: $Ax + By + Cz = D$	240
Intersection of Two Lines in $\mathbb{R}^3$: Parallel/Coincident Case	243
Intersection of Two Lines in $\mathbb{R}^3$: Non-Parallel/Non-Coincident Case	244
Intersection of Two Planes	245
Intersection of Three Planes: Parallel/Coincident Case	246
Intersection of Three Planes: Non-Parallel/Non-Coincident Case	249

A Few Terms in Sequences and Series and a Sampling of Statistics

Summation Notation and Common SUM $\equiv \sum$ Formulas	250
Arithmetic and Geometric Sequences and Series	251
Combinations and Permutations: Choosing and Arranging	252
Elementary Probability	255
Mean, Median, Mode, and Standard Deviation	256
The Binomial Theorem	257
Proof by Induction	258

Index	261
Almost Every Integration Formula You'll Ever Need!	266
Trig Ratio Formula Page	268

How to... How to... How to...

Get an "A" in MATH!

1 After class, **DON'T** do your homework! Instead, *read over your class notes*. When you come to an example done in class...

2 **DON'T** read the example.

Copy out the question, set your notes aside, and do the question yourself. Maybe you will get stuck. Even if you thought you understood the example completely when the teacher went over it in class, you may get stuck.

And this is **GOOD NEWS!** Now, you know what you don't know. So, consult your notes, look in the text, see your teacher/professor. Do whatever is necessary to figure out the steps in the example that troubled you.

Once you have sweated through the example, **DO IT AGAIN! And again**. Do it as often as you need so that it becomes, if not easy, then at least straightforward. Make sure you not only understand each line in the solution, but why each line is needed for the solution.

In part, you have memorized the solution. More importantly, you have made the subtleties of the problem unsubtle!

This is the great equalizer step. If your math or science aptitude is strong, then maybe you will have the example down pat after doing it twice. If not so strong, you may have to do it several times. But after you have done this for every class example...

3 **DO** YOUR HOMEWORK!

If you follow this method and if the teacher chose the examples well, then most of the homework questions will relate easily back to problems done in class and the rest should extend or synthesize the ideas behind those problems.

Guess what you'll find on 80% or more of your tests and exams? The same kinds of problems! And you will have your "A". Good luck, although if you use this method, luck will have nothing to do with your INEVITABLE success.

Copyright © 2017 by Nelson Education Ltd.

How to... How to... How to... Get Extra Help

1 **Have your questions ready.** When you see your teacher for extra help, **don't say anything like**, "I don't have a clue what's going on." Rather, work through your class notes – definitions, examples, theorems, and proofs – thoroughly, and be prepared to say, "I understand everything up to this point. How did we get from here to here?" In other words, **do your part!** Spend quality time with the material.

2 **Get into study groups.** Then one representative of your group can see your teacher for help with problems and report back to the others. Also, with group expertise, more often than not, you will solve most problems yourselves.

3 **Struggle more than a little.** Don't give up after one attempt. Make a sincere effort to sort out your problems. That way, when you say, "I am stuck RIGHT HERE!", you will be so **up** on the problem that your teacher's explanation will be clear.

Have you ever had this experience?
A teacher is explaining a concept or technique in answer to your question. You are nodding your head, saying, "Uh huh! Uh huh! Yes, I understand." Yet you're thinking, "I don't have a clue what the teacher is talking about."

It can happen, but it's **rare**, that the teacher isn't explaining the problem well. Usually, though, the blame lies, yes, with the student, who hasn't struggled enough so that the teacher's explanation can work. **Struggle more than a little!**

- Review each day's notes as soon as possible – definitely before the next class.
- Do your homework before the next class.
- Participate in class.
- Form study groups with classmates.
- Don't fall behind.

SOME SPECIFIC STRATEGIES

BEDMAS (Order of Operations)

BEDMAS stands for **B** [Brackets] **E** [Exponents] **D** [Division] **M** [Multiplication] **A** [Addition] **S** [Subtraction] and is a mnemonic (a fancy word that means "easy memory trick") for the order in which we do arithmetic operations. So, **B**rackets (or parentheses) come first, then **E**xponents, then **D**ivision and **M**ultiplication **in the order they occur from left to right**, and finally, **A**ddition and **S**ubtraction, **in the order they occur from left to right.**

Example 1) Evaluate (a) $4 \times 3 + 2$ (b) $2 + 4 \times 3$ (c) $(2+4) \times 3$ (d) 2×4^3

Solution (a) $4 \times 3 + 2$ [first × then +] $= 12 + 2 = 14$ (b) $2 + 4 \times 3$ [first × then +] $= 2 + 12 = 14$

(c) $(2+4) \times 3$ [first brackets then ×] $= 6 \times 3 = 18$ (d) 2×4^3 [first exponents then ×] $= 2 \times 64 = 128$

Example 2) Evaluate

(a) $6 \div 3 \times 5$ (b) $6 \div (3 \times 5)$ (c) $6 \times 3 \div 5$ (d) $6 \div 3 \div 5$ (e) $6 \div (3 \div 5)$

Solution (a) $6 \div 3 \times 5$ [first ÷ then ×] $= 2 \times 5 = 10$ (b) $6 \div (3 \times 5)$ [first ( ) then ÷] $= 6 \div 15 = \dfrac{6}{15} = \dfrac{2}{5}$

Note: In (a), BEDMAS had us divide first. In (b), with brackets, we first did the multiplication. Different answers!

(c) $6 \times 3 \div 5$ [first × then ÷] $= 18 \div 5 = \dfrac{18}{5}$

... the answer is different again. BEDMAS is our friend. It brings (arithmetic) *order* to our lives!

BEDMAS

(d) $6 \div 3 \div 5$ [Do the **first** division first.] $= 2 \div 5 = \dfrac{2}{5}$

Note: Suppose in (d), we did the second division first. THIS IS WRONG!

(d) done incorrectly : $6 \div 3 \div 5$ [Do the **second** division first. THIS IS WRONG!] $= 6 \div \dfrac{3}{5}$ [Invert and multiply.] $= \overset{2}{\cancel{6}} \times \dfrac{5}{\underset{1}{\cancel{3}}} = 10$. Wrong!

The order of the ÷ signs determines the order in which they are carried out. However, ...

(e) $6 \div (3 \div 5)$ [Do the second division first. This is **correct** because of the brackets.] $= 6 \div \dfrac{3}{5}$ [Invert and multiply.] $= \overset{2}{\cancel{6}} \times \dfrac{5}{\underset{1}{\cancel{3}}} = 10$

Adding and Subtracting Fractions

Wouldn't it be great if " $\dfrac{a}{b} + \dfrac{c}{d} = \dfrac{a+c}{b+d}$ " [**This is wrong!**], that is, when adding fractions, we just added the numerators and denominators?

Well, sorry, life is not that simple. But life is not that hard either. When you add or subtract fractions, you need a **common denominator**. Better still, you should use the **lowest common denominator**, as you will see. However, this method always works:

[**This is right!**]
$$\frac{a}{b} + \frac{c}{d} = \frac{a \times d}{b \times d} + \frac{c \times b}{d \times b} = \frac{ad + bc}{bd}$$

Example 1) (a) $\dfrac{1}{3} + \dfrac{5}{7}$ (b) $1\dfrac{5}{6} - 2\dfrac{2}{5}$ (c) $\dfrac{5}{9} + \dfrac{5}{12}$

Solution (a) $\dfrac{1}{3} + \dfrac{5}{7}$ [The common denominator is $3 \times 7 = 21$.] $= \dfrac{1 \times 7}{3 \times 7} + \dfrac{5 \times 3}{7 \times 3} = \dfrac{7 + 15}{21} = \dfrac{22}{21}$ [and as a mixed fraction...] $= 1\dfrac{1}{21}$

(b) $1\dfrac{5}{6} - 2\dfrac{2}{5}$ [Make improper fractions.] $= \dfrac{11}{6} - \dfrac{12}{5} =$ [The common denominator is $6 \times 5 = 30$.] $= \dfrac{55}{30} - \dfrac{72}{30} = -\dfrac{17}{30}$

(c) $\dfrac{5}{9} + \dfrac{5}{12}$ [The **smallest** common denominator is $9 \times 4 = 36$ which is the **least common multiple** of 9 and 12!] $= \dfrac{5 \times 4}{9 \times 4} + \dfrac{5 \times 3}{12 \times 3} = \dfrac{20}{36} + \dfrac{15}{36} = \dfrac{35}{36}$

Example 2) (a) $\dfrac{1}{3} + \dfrac{1}{5} - \dfrac{1}{7}$ (b) $\dfrac{2}{15} + \dfrac{3}{10} - \dfrac{1}{6} + 3$

Solution (a) $\dfrac{1}{3} + \dfrac{1}{5} - \dfrac{1}{7}$ [The common denominator is $3 \times 5 \times 7 = 105$.] $= \dfrac{1 \times 5 \times 7}{105} + \dfrac{1 \times 3 \times 7}{105} - \dfrac{1 \times 3 \times 5}{105} = \dfrac{35 + 21 - 15}{105} = \dfrac{41}{105}$

(b) $\dfrac{2}{15} + \dfrac{3}{10} - \dfrac{1}{6} + 3 = \dfrac{2}{3 \times 5} + \dfrac{3}{2 \times 5} - \dfrac{1}{2 \times 3} + \dfrac{3}{1}$ [The **lowest** common denominator is $3 \times 5 \times 2 = 30$.] $= \dfrac{4}{30} + \dfrac{9}{30} - \dfrac{5}{30} + \dfrac{90}{30} = \dfrac{98}{30} = \dfrac{49}{15}$

Note: See page 22 for adding and subtracting polynomial fractions.

Multiplying and Dividing Fractions

When multiplying fractions, life is easy: $\boxed{\dfrac{a}{b} \times \dfrac{c}{d} = \dfrac{ac}{bd}}$

Dividing fractions causes a little heartache, but only because many of you don't understand the famous **"invert and multiply"** rule.

If you have divided five **whole** pizzas into thirds, then certainly you have fifteen portions. Well, you have just verified that $\dfrac{5 \text{ pizzas}}{\left(\dfrac{1}{3}\right)} = 15$ pizza portions, that is, you have inverted

and multiplied: $\dfrac{5}{\left(\dfrac{1}{3}\right)} = 5 \times \overset{\text{Invert and multiply!}}{\left(\dfrac{3}{1}\right)} = 15$

Example 1) (a) $\dfrac{4}{5} \times \dfrac{3}{7}$ (b) $\dfrac{14}{9} \times \dfrac{3}{7}$ (c) $2\dfrac{2}{5} \times 3\dfrac{1}{4}$

Solution (a) $\dfrac{4}{5} \times \dfrac{3}{7} \overset{\frac{a}{b} \times \frac{c}{d} = \frac{ac}{bd}}{=} \dfrac{12}{35}$ (b) $\dfrac{14}{9} \times \dfrac{3}{7} \overset{\text{Reduce first.}}{=} \dfrac{\overset{2}{\cancel{14}}}{\underset{3}{\cancel{9}}} \times \dfrac{\overset{1}{\cancel{3}}}{\underset{1}{\cancel{7}}} = \dfrac{2}{3} \times \dfrac{1}{1} = \dfrac{2}{3}$

(c) $2\dfrac{2}{5} \times 3\dfrac{1}{4} \overset{\text{Make improper fractions.}}{=} \dfrac{\overset{3}{\cancel{12}}}{5} \times \dfrac{13}{\underset{1}{\cancel{4}}} = \dfrac{39}{5} \overset{\text{or, as a mixed fraction, is ...}}{=} 7\dfrac{4}{5}$

Example 2) (a) $\dfrac{\left(\dfrac{4}{5}\right)}{\left(\dfrac{3}{7}\right)}$ (b) $\dfrac{\left(\dfrac{7}{3}\right)}{4}$ (c) $\dfrac{7}{\left(\dfrac{3}{4}\right)}$

Solution (a) $\dfrac{\left(\dfrac{4}{5}\right)}{\left(\dfrac{3}{7}\right)} \overset{\text{Invert and multiply!}}{=} \dfrac{4}{5} \times \dfrac{7}{3} = \dfrac{28}{15}$ (b) $\dfrac{\left(\dfrac{7}{3}\right)}{4} = \dfrac{\left(\dfrac{7}{3}\right)}{\left(\dfrac{4}{1}\right)} = \dfrac{7}{3} \times \dfrac{1}{4} = \dfrac{7}{12}$

(c) $\dfrac{7}{\left(\dfrac{3}{4}\right)} = \dfrac{\left(\dfrac{7}{1}\right)}{\left(\dfrac{3}{4}\right)} = \dfrac{7}{1} \times \dfrac{4}{3} = \dfrac{28}{3}$

Operations with Decimals

If I ask you to evaluate $\frac{50}{5}$, I **know** you are going to tell me the answer is 10. However, if I ask you to evaluate $\frac{50}{.05}$, many of you are going to search frantically for your calculator. No calculator needed: $\frac{50}{.05} = \frac{50 \cdot 100}{.05 \cdot 100} = \frac{5000}{5} = 1000$

Example 1) When adding or subtracting with decimals, align the decimal points.
Evaluate: $4.23 + 0.045$

For addition, align the decimals:
$$4.230 + 0.045$$

Solution $4.23 + 0.045 = 4.275$

Example 2) When subtracting, if the number you are subtracting has the larger magnitude, you may want to "turn the question around".
Evaluate: $3.7 - 15.02$

Turn this around! Note the minus sign.

Solution $3.7 - 15.02 = -(15.02 - 3.7) = -11.32$

Example 3) When multiplying, count the number of decimals in the question to determine the number of decimal places for the answer.
Evaluate: $.0003 \times 4.2$

First, $3 \times 42 = 126$. We need 5 decimal places in the answer.

Solution $.0003 \times 4.2 = .00126$

Example 4) When dividing, multiply the numerator and denominator by a power of 10 so that both the top and bottom become integers.
Evalutate: $\frac{1.60}{.0004}$

Multiply top and bottom by the power of 10 that converts all numbers to integers.

Solution $\frac{1.60}{.0004} = \frac{1.6 \cdot 10000}{.0004 \cdot 10000} = \frac{16000}{4} = 4000$

Of course, I cooked the numbers in these examples so that a calculator really wasn't necessary. Most "real" examples will turn into time wasters without the calculator. So as long as you **understand** how decimals work, calculate away!

Two For You – BEDMAS (Order of Operations)

1)(a) Evaluate: $22 + 14 \div 3 \times 4 - 3$

(b) The question in (a) was sent to me by the mom of a student who had won second prize in a Honda contest—$200 in CD's. First prize was a Honda motorcycle. Mom was relieved her son won second prize! He submitted the **correct** answer to the "skill testing question". However, Honda claimed the answer was 45 and withheld the prize.
(i) How did Honda arrive at 45?
(ii) Insert brackets into the expression so that the answer is 45.

(c) Mom asked me to write a letter to Honda explaining why her son's answer was correct. I did so. The good news: Honda agreed and gave him the prize. However, I told Mom my time was valuable and that, while I was glad to help, I had a consultation fee. She was to pay me $\$1000 - \$500 \times 2$.
(i) How much did she owe me using Honda's method?
(ii) How much did she owe me using BEDMAS?

Postscript: Mom told me a cheque was in the mail. I'm still waiting!

2) Evaluate: (a) $\dfrac{\left(\dfrac{2}{3}\right)}{\left(\dfrac{4}{5}\right)}$ (b) $\dfrac{\left(\dfrac{7}{5}\right)}{5}$ (c) $\dfrac{7}{\left(\dfrac{4}{5}\right)}$

Answers 1)(a) $37\dfrac{2}{3}$ (b)(i) Honda did each operation in the order in which it appeared.
(b)(ii) $(22+14) \div 3 \times 4 - 3$ (c)(i) $1000   (ii) $0
2)(a) $\dfrac{5}{6}$ (b) $\dfrac{7}{25}$ (c) $\dfrac{35}{4} = 8\dfrac{3}{4}$

Two For You – Adding and Subtracting Fractions

1)(a) $\dfrac{1}{5} + \dfrac{2}{9}$ (b) $3\dfrac{5}{7} - 4\dfrac{1}{3}$ (c) $\dfrac{2}{11} + \dfrac{1}{22}$

2)(a) $\dfrac{1}{2} + \dfrac{2}{5} - 1\dfrac{1}{10}$ (b) $5 - 2\dfrac{1}{4} + \dfrac{2}{3}$

Answers 1)(a) $\dfrac{19}{45}$ (b) $-\dfrac{13}{21}$ (c) $\dfrac{5}{22}$

2)(a) $-\dfrac{1}{5}$ (b) $\dfrac{41}{12}$ or $3\dfrac{5}{12}$

Two For You – Multiplying and Dividing Fractions

1) (a) $\dfrac{2}{9} \times \dfrac{4}{5}$ (b) $\dfrac{4}{5} \times \dfrac{15}{16}$ (c) $1\dfrac{1}{3} \times 2\dfrac{2}{7}$

2) (a) $\dfrac{\left(\dfrac{2}{9}\right)}{\left(\dfrac{5}{8}\right)}$ (b) $\dfrac{\left(\dfrac{2}{11}\right)}{3}$ (c) $\dfrac{5}{\left(\dfrac{1}{7}\right)}$

Answers 1)(a) $\dfrac{8}{45}$ (b) $\dfrac{3}{4}$ (c) $\dfrac{64}{21}$ or $3\dfrac{1}{21}$

2)(a) $\dfrac{16}{45}$ (b) $\dfrac{2}{33}$ (c) $\dfrac{35}{1} = 35$

Two For You – Operations with Decimals

Evaluate using pen and paper!

1) (a) $20.415 + .017 + .00001$ (b) $9.082 - 5.61$ (c) $5.61 - 9.082$

2) (a) $4.5 \times .02 \times .0003$ (b) $\dfrac{.0025}{500}$

Answers 1)(a) 20.43201 (b) 3.472 (c) -3.472

2)(a) $.000027$ (b) $.000005$

Ratio, Rate, and Percent

A ratio is a comparison of relative size between two numbers. If I said the ratio of males to females in a class was 2:3 you know there could more than 5 students in the room. If I told you there were 10 guys in the room, I hope you could tell me there must be 15 girls.

Example 1) Solve for the unknown. (a) $3:4 = 15:x$ (b) $6:18 = 9:b$

Solution (a)

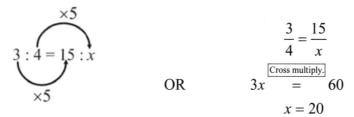

$$\frac{3}{4} = \frac{15}{x}$$
Cross multiply.
OR $3x = 60$
$x = 20$

Therefore, $x = 20$.

(b)

$$\frac{6}{18} = \frac{9}{b}$$
Cross multiply.
OR $6b = 162$
$b = 27$

Therefore, $b = 27$.

A percent is a ratio where the comparison is always to 100. For example, 7 out of 10 is equivalent to 70 out of 100, that is 70%. You can represent a percent as a decimal by dividing by 100; for example, $43\% = \frac{43}{100} = 0.43$.

Example 2) A $500 TV is on sale for 35% off. What is the sale price of the TV?
Solution The discount is $0.35 \times \$500 = \$175$, so the sale price of the TV is $\$500 - \$175 = \$325$.

A rate is a ratio where the comparison is always to 1. If you drive 400 km in 4 hours, then your rate is 100 km per 1 hour, that is, 100 km/h.

Example 3) John spent $87.50 to see 7 movies in theatres and he plans to see four more movies. How much money will he need? (Assume the admission price is always the same.)

Solution The rate is $\$87.50 \div 7 = \$12.50$ per movie. This means John needs $\$12.50 \times 4 = \$50$ to see four more films.

Complex Numbers

When solving for the roots of a quadratic relationship $y = ax^2 + bx + c$, you use the quadratic formula, $x = \dfrac{-b \pm \sqrt{b^2 - 4ac}}{2a}$. The sign of $b^2 - 4ac$ tells you a lot about the roots of the quadratic.

	Number of real roots	$a > 0$	$a < 0$
$b^2 - 4ac > 0$	2		
$b^2 - 4ac = 0$	1		
$b^2 - 4ac < 0$	0		

The issue when $b^2 - 4ac < 0$ is that we need to take the square root of a negative number, which we cannot do when working with real numbers. The square of any real number is positive! There are still roots to the quadratic; we just need to work in the "complex number system". The complex number system is formed by including a new number to deal with square roots of negatives. We define $i = \sqrt{-1}$. This allows us to define a way to handle the square root of a negative number (as well as 4th roots and 6th roots and 8th roots, etc.). For example, we define $\sqrt{-81}$ to be $\sqrt{81} \times \sqrt{-1} = 9i$. The letter i stands for "imaginary". This may seem like mathematicians playing some sort of weird game, but if you think about it, all numbers are just concepts. In fact, i is just as "real" as 3 or $\dfrac{1}{2}$ or $\sqrt{2}$ or π or … Complex numbers can be added, subtracted, multiplied and divided just like any other number. Treat i as you would treat a variable like x and keep in mind that $i^2 = -1$.

Example 1) Simplify: a) $3i + 4i - 9i$ b) $(3+i)(4-i)$ c) $\dfrac{4+2i}{2i}$ d) i^3

Solution a) $3i + 4i - 9i = 7i - 9i = -2i$

b) $(3+i)(4-i) = 12 - 3i + 4i - i^2 = 12 + i - (-1) = 13 + i$

c) $\dfrac{4+2i}{2i}$ [Multiply top and bottom by i.] $= \dfrac{4i + 2i^2}{2i^2}$ [$i^2 = -1$] $= \dfrac{4i - 2}{-2} = -2i + 1$

d) $i^3 = i^2 i = -i$

Factoring Difference of Squares

Does anyone out there have a problem with $a^2 - b^2 = (a-b)(a+b)$? I didn't think so! Problems arise when it's not so obvious that difference of squares is what we are dealing with.

Example 1) Factor: (a) $a^4 - b^8$ (b) $(x+y-z)^2 - (x-y-z)^2$

Solution (a) $a^4 - b^8 = (a^2-b^4)(a^2+b^4) \stackrel{\text{Don't stop now!}}{=} (a-b^2)(a+b^2)(a^2+b^4)$

$\stackrel{\text{for the obsessive}}{=} (\sqrt{a}-b)(\sqrt{a}+b)(a+b^2)(a^2+b^4)$

(b) $(x+y-z)^2 - (x-y-z)^2 \stackrel{\substack{a=x+y-z \\ b=x-y-z}}{=} \big(x+y-z-(x-y-z)\big)\big(x+y-z+x-y-z\big)$
$= 2y(2x-2z) = 4y(x-z)$

Difference of squares is often used **BACKWARDS**, to rationalize expressions. This is especially useful in some limit questions.

Example 2) Rationalize the denominator in $\dfrac{1}{\sqrt{x}+\sqrt{y}}$.

Solution $\dfrac{1}{\sqrt{x}+\sqrt{y}} = \dfrac{1}{\sqrt{x}+\sqrt{y}}\left(\dfrac{\sqrt{x}-\sqrt{y}}{\sqrt{x}-\sqrt{y}}\right) = \dfrac{\sqrt{x}-\sqrt{y}}{x-y}$

Example 3) Evaluate: $\lim\limits_{x\to 9}\dfrac{x-9}{\sqrt{x}-3}$

First Solution $\lim\limits_{x\to 9}\dfrac{x-9}{\sqrt{x}-3} \stackrel{\text{Factor the top...}}{=} \lim\limits_{x\to 9}\dfrac{(\sqrt{x}-3)(\sqrt{x}+3)}{\sqrt{x}-3} = \lim\limits_{x\to 9}(\sqrt{x}+3) = 6$

Second Solution $\lim\limits_{x\to 9}\dfrac{x-9}{\sqrt{x}-3} \stackrel{\text{...or rationalize the bottom.}}{=} \lim\limits_{x\to 9}\dfrac{x-9}{(\sqrt{x}-3)}\left(\dfrac{\sqrt{x}+3}{\sqrt{x}+3}\right)$

$= \lim\limits_{x\to 9}\dfrac{(x-9)(\sqrt{x}+3)}{x-9} = \lim\limits_{x\to 9}(\sqrt{x}+3) = 6$

Factoring Difference of Cubes

Lots of people have trouble with this one!
$a^3 - b^3 = (a-b)(a^2 + ab + b^2)$. Note that the coefficient of ab is +1. Note also that $a^2 + ab + b^2$ doesn't factor any further!

Example 1) Factor: (a) $a^3 - b^6$ (b) $(x+h)^3 - x^3$

Solution (a) $a^3 - b^6 \overset{a^3-(b^2)^3}{=} (a-b^2)(a^2 + ab^2 + b^4)$

$\overset{\text{For the obsessive: use difference of squares.}}{=} (\sqrt{a} - b)(\sqrt{a} + b)(a^2 + ab^2 + b^4)$

(b) $(x+h)^3 - x^3 \overset{\text{Here, } a=x+h \text{ and } b=x.}{=} (x+h-x)\big((x+h)^2 + (x+h)x + x^2\big)$

$= h\big((x+h)^2 + (x+h)x + x^2\big)$

(If you have taken calculus, this example should remind you of the derivative limit for $y = x^3$.) As with difference of squares, there are two approaches to limits with difference of cubes.

Example 2) Evaluate $\lim\limits_{x \to 64} \dfrac{x - 64}{x^{1/3} - 4}$.

First Solution $\lim\limits_{x \to 64} \dfrac{x - 64}{x^{1/3} - 4} \overset{\text{Factor the top: } a=x^{1/3} \text{ and } b=4.}{=} \lim\limits_{x \to 64} \dfrac{(x^{1/3} - 4)(x^{2/3} + 4x^{1/3} + 16)}{x^{1/3} - 4}$

$= \lim\limits_{x \to 64}(x^{2/3} + 4x^{1/3} + 16) = 48$

Second solution $\lim\limits_{x \to 64} \dfrac{x - 64}{x^{1/3} - 4} \overset{\text{Now rationalize the bottom.}}{=} \lim\limits_{x \to 64} \dfrac{x - 64}{(x^{1/3} - 4)} \left(\dfrac{x^{2/3} + 4x^{1/3} + 16}{x^{2/3} + 4x^{1/3} + 16} \right)$

$= \lim\limits_{x \to 64} \dfrac{(x - 64)(x^{2/3} + 4x^{1/3} + 16)}{x - 64} = \lim\limits_{x \to 64}(x^{2/3} + 4x^{1/3} + 16) = 48$

Three For You – Ratio, Rate and Percent

1) Solve for the unknown variables.

a) $4:3 = 16:a$ b) $2:b = 7:14$ c) $2:3 = 6:c$

2)(a) Jack bought 4 cases of pop for $15. Gord bought 6 cases for $19.98. Who got the better deal?

(b) A car can drive 400 km on 36 L of gas. How many litres will you need to drive 645 km?

3) Regular price for a pair of shoes is $60. If the shoes are on sale for 30% off and tax is 13%, what is the price of the shoes after tax?

Answers 1)(a) $a = 12$ (b) $b = 4$ (c) $c = 9$
2) (a) Gord ($3.33 per case versus $3.75 per case)   (b) 58.05 L
3) $47.46 (Discount = $18, Sale price = $42, Tax = $5.46)

Two For You – Complex Numbers

1) Simplify: (a) $(7+i)(6-i)$ (b) $\dfrac{7-i}{3+i}$ Hint: multiply by $\dfrac{3-i}{3-i}$.   (c) i^9

2) Find the roots of $x^2 + 4x + 5$ using the quadratic formula.

Answers 1)(a) $43 - i$ (b) $2 - i$ (c) i
2) $x = -2 + i$ and $x = -2 - i$

Two For You – Factoring Difference of Squares

1) Factor: (a) $(x+y)^2 - (x-y)^2$ (b) $x^4 - z^{12}$

2)(a) Rationalize the numerator: $\dfrac{\sqrt{x+4} - \sqrt{3x-6}}{x-5}$

(b) Evaluate: $\lim\limits_{x \to 5} \dfrac{\sqrt{x+4} - \sqrt{3x-6}}{x-5}$

Answers 1)(a) $4xy$ (b) $(x-z^3)(x+z^3)(x^2+z^6)$

2)(a) $\dfrac{-2}{\sqrt{x+4} + \sqrt{3x-6}}$ (b) $-\dfrac{1}{3}$

Two For You – Factoring Difference of Cubes

1) Factor: $x^6 - y^6$

2) Evaluate the limits: (a) $\lim\limits_{x \to 5} \dfrac{x^3 - 125}{x^2 - 25}$ (b) $\lim\limits_{x \to 8} \dfrac{x-8}{x^{1/3} - 2}$

Answers 1) $(x^2 - y^2)(x^4 + x^2y^2 + y^4) = (x-y)(x+y)(x^4 + x^2y^2 + y^4)$

2)(a) $\dfrac{15}{2}$ (b) 12

Factoring $a^n - b^n$ and $a^n + b^n$

First: Factoring $a^n - b^n$

First, please review these topics:
Difference of Squares (page 11) and **Difference of Cubes** (page 12).

$$a^2 - b^2 = (a-b)(a+b)$$
$$a^3 - b^3 = (a-b)(a^2 + ab + b^2) \text{ Note that the coefficient of } ab \text{ is 1.}$$
$$a^4 - b^4 = (a-b)(a^3 + a^2b + ab^2 + b^3). \text{ So, for positive integers } n,$$
$$a^n - b^n = (a-b)(a^{n-1} + a^{n-2}b + a^{n-3}b^2 + a^{n-4}b^3 + \ldots + a^2b^{n-3} + ab^{n-2} + b^{n-1})$$

In the second bracket for the factored form of $a^n - b^n$, the exponent on a starts at $n-1$ and decreases one by one down to 0. The exponent on b starts at 0 and goes up one by one to $n-1$. **This formula works for any $n \in \mathbb{N}$**, that is, for **natural numbers**.

Second: Factoring $a^n + b^n$ (and we want n to be ODD!)

Face it: $a^2 + b^2$ **doesn't factor!** Well, all right, it does if you allow **complex numbers**. $a^2 + b^2 = (a - bi)(a + bi)$, where $i = \sqrt{-1}$. For our purposes, restricted to real numbers, the sum of squares doesn't factor. Also, $a^4 + b^4$ and $a^8 + b^8$ have no easy linear factors like $a - b$ or $a + b$ (although they do have complicated quadratic factors! For now, don't ask!) However, sum of cubes does factor: $a^3 + b^3 = (a+b)(a^2 - ab + b^2)$
Compare the difference of cubes: $a^3 - b^3 = (a-b)(a^2 + ab + b^2)$
Look at where + changes to −. So, as long as n is $\boxed{\text{ODD}}$,
$$a^n + b^n = (a+b)(a^{n-1} - a^{n-2}b + a^{n-3}b^2 - a^{n-4}b^3 + \ldots + a^2b^{n-3} - ab^{n-2} + b^{n-1})$$

Example 1) Factor: (a) $a^5 - b^5$ (b) $a^5 + b^5$ (c) $a^7 + b^7$

Solution (a) $a^5 - b^5 = (a-b)(a^4 + a^3b + a^2b^2 + ab^3 + b^4)$
(b) $a^5 + b^5 = (a+b)(a^4 - a^3b + a^2b^2 - ab^3 + b^4)$
(c) $a^7 + b^7 = (a+b)(a^6 - a^5b + a^4b^2 - a^3b^3 + a^2b^4 - ab^5 + b^6)$

Common Factors

The easiest kind of factoring is "common factoring". However, even common factors can be confusing when terms have factors like $(a+b)$ or negative exponents or fractional exponents! (expo-nents!) Let's do basics first and challenges second. Remember:
$ax + ay - az = a(x + y - z)$.

Example 1) Factor: (a) $2x^4y^2 + 4x^3y^3$ (b) $4(a+b)^2 - 7(a+b)^3 + a + b$

Solution (a) $2x^4y^2 + 4x^3y^3 = 2x^3y^2(x + 2y)$
[$2x^3y^2$ is common to both terms.]

(b) $4(a+b)^2 - 7(a+b)^3 + a + b = (a+b)\left(4(a+b) - 7(a+b)^2 + 1\right)$
[$(a+b)$ is common to all three terms, treating the last $(a+b)$ as a single term.]

Example 2) Use common factoring to simplify the following:

(a) $\dfrac{(x-1)^3(3)(x+1)^2 - (x+1)^3(3)(x-1)^2}{(x-1)^6}$ (b) $x^{\frac{1}{3}}\left(\dfrac{2}{3}\right)(x-5)^{-\frac{1}{3}} + (x-5)^{\frac{2}{3}}\left(\dfrac{1}{3}\right)x^{-\frac{2}{3}}$

> Calculus students: these expressions appear with $\dfrac{d}{dx}\left(\dfrac{(x+1)^3}{(x-1)^3}\right)$ and $\dfrac{d}{dx}\left(x^{\frac{1}{3}}(x-5)^{\frac{2}{3}}\right)$.

Solution (a) $\dfrac{(x-1)^3(3)(x+1)^2 - (x+1)^3(3)(x-1)^2}{(x-1)^6}$

[$3(x-1)^2(x+1)^2$ is common to both terms in the numerator.]

$= \dfrac{3(x-1)^2(x+1)^2}{(x-1)^6}\left((x-1) - (x+1)\right) = -\dfrac{6(x+1)^2}{(x-1)^4}$

[The common factor is $\left(\dfrac{1}{3}\right)x^{-\frac{2}{3}}(x-5)^{-\frac{1}{3}}$. The **LOWER** exponent is the common exponent!]

(b) $x^{\frac{1}{3}}\left(\dfrac{2}{3}\right)(x-5)^{-\frac{1}{3}} + (x-5)^{\frac{2}{3}}\left(\dfrac{1}{3}\right)x^{-\frac{2}{3}}$

[I am including this optional step to show **EXPLICITLY** the common factor with the negative exponent in each of the first and second terms.]

[This is $x^{\frac{1}{3}}$ rewritten as $x^{-\frac{2}{3}}x^{\frac{3}{3}} = x^{-\frac{2}{3}}x$ in the first term to emphasize how to factor out $x^{-\frac{2}{3}}$ from this term.]

[This is $(x-5)^{\frac{2}{3}}$ rewritten as $(x-5)^{-\frac{1}{3}}(x-5)^{\frac{3}{3}} = (x-5)^{-\frac{1}{3}}(x-5)$ in the second term to emphasize how to factor $(x-5)^{-\frac{1}{3}}$ from this term.]

$= x^{-\frac{2}{3}}x \cdot \left(\dfrac{2}{3}\right)(x-5)^{-\frac{1}{3}} + (x-5)^{-\frac{1}{3}}(x-5) \cdot \left(\dfrac{1}{3}\right)x^{-\frac{2}{3}}$

$= \left(\dfrac{1}{3}\right)x^{-\frac{2}{3}}(x-5)^{-\frac{1}{3}}\left(\underbrace{2x}_{\text{first term}} + \underbrace{x-5}_{\text{second term}}\right)$

[This is all that is left in the first term after taking out the common factor.] [This is all that is left in the second term after taking out the common factor.]

$= \left(\dfrac{1}{3}\right)x^{-\frac{2}{3}}(x-5)^{-\frac{1}{3}}(3x-5) = \dfrac{3x-5}{3x^{\frac{2}{3}}(x-5)^{\frac{1}{3}}}$

Factoring Quadratic Expressions/Trinomials Without Using the Quadratic Formula

Some quadratic expressions factor easily: $x^2 - 4x + 3 = (x-3)(x-1)$

When we need to factor quadratic expressions with equations like $2x^2 - x - 5 = 0$ (which in fact has **non-rational roots**) or $2x^2 - x + 3 = 0$ (which in fact has **non-real roots**), we can use the quadratic formula to make factoring foolproof. Go to page 53 for that method (and to see why the roots are non-rational and non-real, respectively.) However, when the quadratic expression/trinomial factors easily, using the quadratic formula is too much work when this simpler approach will do.

Assume a and b are positive and $a \leq b$.

This is crucial to understanding the method below. There are four scenarios in the...

TRINOMIAL BOX

(1) $(x-a)(x-b) = x^2 - ax - bx + ab = x^2 - (a+b)x + ab$
(2) $(x+a)(x+b) = x^2 + ax + bx + ab = x^2 + (a+b)x + ab$
(3) $(x-a)(x+b) = x^2 - ax + bx - ab = x^2 + (b-a)x - ab$
(4) $(x+a)(x-b) = x^2 + ax - bx - ab = x^2 - (b-a)x - ab$

Example 1) Factor $x^2 - 4x + 3$.

Solution This trinomial fits the form (1) in the trinomial box above: $a + b = 4$ and $ab = 3$.

$x^2 - 4x + 3 \overset{\substack{a=1,\,b=3 \\ (x-a)(x-b)}}{=} (x-1)(x-3)$.

Example 2) Factor: (i) $x^2 + 4x - 5$ (ii) $x^2 - 4x - 5$

Solution (i) Because of the -5 and the $+4$, we have form (3) in the Trinomial Box.

$x^2 + 4x - 5 \overset{\substack{a=1,\,b=5 \\ (x-a)(x+b)}}{=} (x-1)(x+5)$

(ii) Because of the -5 and the -4, we have form (4) in the Trinomial Box.

$x^2 - 4x - 5 \overset{\substack{a=1,\,b=5 \\ (x+a)(x-b)}}{=} (x+1)(x-5)$

Example 3) Factor: $3x^2 + 14x - 5$

Solution This doesn't fit into the Trinomial Box. It does factor pretty easily but we need a fair bit of trial and error to make the x^2 coefficient 3, the constant -5 and the x coefficient 14. My suggestion: it is easy and "trial and error" free to use the quadratic formula. Go to page 53.

The Remainder and Factor Theorems for Polynomials

The Remainder Theorem tells us that when we divide polynomial $P(x)$ by $x-a$ to obtain quotient $Q(x)$ and remainder R, then $R = P(a)$, that is, $P(x) = Q(x)(x-a) + P(a)$. So, if $P(a) = 0$, then $P(x) = Q(x)(x-a)$, that is, $x-a$ is a factor of $P(x)$. This is ...

> **The Factor Theorem**
> $x-a$ is a factor of polynomial $P(x) \Leftrightarrow P(a) = 0$.

Also, suppose $P(x) = a_n x^n + a_{n-1} x^{n-1} + \ldots + a_1 x + a_0$, where all the a_i's are integers, and $\frac{p}{q}$ is a root of $P(x)$. ***Then q must divide a_n and p must divide a_0.***

Example 1) Using the Factor Theorem, find **rational roots** of $P(x) = x^3 - x^2 - 4x + 4$.

Solution The coefficient of x^3 is 1 and $a_0 = 4$. If $a = \frac{p}{q}$ is a rational root, then p must divide 4 and q must divide 1. Therefore, we need to check $a = \pm 1, \pm 2, \pm 4$:
$P(1) = 0, \; P(-1) = 6, \; P(2) = 0, \; P(-2) = 0, \; P(4) = 36, \; P(-4) = -60$.
Therefore the roots are 1, 2, and -2.

Two notes : 1) $P(x)$, a cubic, can have at most 3 roots. If we find 3 roots, we are done!
   2) $P(x) = (x-1)(x-2)(x+2)$

Example 2) Find the rational roots of $P(x) = 2x^4 - 5x^3 + 5x^2 - 5x + 3$.

Solution The coefficient of x^4 is 2 and $a_0 = 3$. If $a = \frac{p}{q}$ is a rational root, then p must divide 3 and q must divide 2. We need to check $a = \pm 1, \pm \frac{1}{2}, \pm 3,$ and $\pm \frac{3}{2}$:

$P(1) = 0, \; P(-1) = 20, \; P(3) = 60, \; P(-3) = 360, \; P\left(\frac{1}{2}\right) = \frac{5}{4}, \; P\left(-\frac{1}{2}\right) = \frac{15}{2}, \; P\left(\frac{3}{2}\right) = 0,$
$P\left(-\frac{3}{2}\right) = \frac{195}{4}$. Therefore the rational roots are 1 and $\frac{3}{2}$.

Three notes : 1) Since $P(x)$ is degree 4, the other roots are irrational or complex.

2) [See Polynomial Division on page 24.] Divide $x-1$ into $P(x)$ and then $x - \frac{3}{2}$ into the resulting quotient.

The result: $P(x) = (x-1)\left(x - \frac{3}{2}\right)(2x^2 + 2) \underset{\text{a little prettier...}}{=} (x-1)(2x-3)(x^2+1)$

3) The non-rational roots are $\pm i$, that is, $\pm\sqrt{-1}$.

Two For You – Factoring $a^n - b^n$ and $a^n + b^n$

1) Factor: $a^5 - b^{10}$
2) Factor: $a^{15} + b^{30}$ (Hint: $a^{15} + b^{30} = (a^3)^5 + (b^6)^5$)

Answers 1) $(a - b^2)(a^4 + a^3 b^2 + a^2 b^4 + ab^6 + b^8)$
2) $(a^3 + b^6)(a^{12} - a^9 b^6 + a^6 b^{12} - a^3 b^{18} + b^{24})$
$\boxed{\text{optional}}$
$= (a + b^2)(a^2 - ab^2 + b^4)(a^{12} - a^9 b^6 + a^6 b^{12} - a^3 b^{18} + b^{24})$

Two For You – Common Factors

Factor each of the following:

1)(a) $3m^3 n^2 - 6m^5 n^5 + 9m^3 n^3 - 12m^4 n^2$ (b) $(x+y)^3 - (x+y)^5 + 2(x+y)$

2) $\left(\dfrac{1}{4}\right)(x-1)^{\frac{1}{2}}(x+1)^{-\frac{3}{4}} + \left(\dfrac{1}{2}\right)(x-1)^{-\frac{1}{2}}(x+1)^{\frac{1}{4}}$

Answers 1)(a) $3m^3 n^2 (1 - 2m^2 n^3 + 3n - 4m)$ (b) $(x+y)\big((x+y)^2 - (x+y)^4 + 2\big)$

2) $\left(\dfrac{1}{4}\right)(x-1)^{-\frac{1}{2}}(x+1)^{-\frac{3}{4}}(3x+1) \boxed{\text{or}} = \dfrac{3x+1}{4(x-1)^{\frac{1}{2}}(x+1)^{\frac{3}{4}}}$

Two For You – Factoring Quadratic Expressions/Trinomials

Factor each of the following:

1)(a) $x^2 + 11x + 10$ (b) $x^2 - 11x + 10$ 2)(a) $x^2 + 11x - 12$ (b) $x^2 - 11x - 12$

Answers 1)(a) $(x+10)(x+1)$ (b) $(x-10)(x-1)$
2)(a) $(x+12)(x-1)$ (b) $(x-12)(x+1)$

Two For You – The Remainder and Factor Theorems

Find the rational roots and then factor.

1) $P(x) = x^3 + 2x^2 - 5x - 6$ 2) $P(x) = 3x^4 - x^3 - 3x + 1$

Answers 1) $-1, 2, -3$, $P(x) = (x+1)(x-2)(x+3)$
2) $1, \frac{1}{3}$, $P(x) = (3x-1)(x-1)(x^2 + x + 1)$

Multiplying Expressions—FOIL: $(a+b)(c+d)$

In grades 9, 10, 11, 12, and yes, even university, there are students who believe $(x+y)^2 = x^2 + y^2$! NoNoNo!!! $(3+4)^2 = 7^2 = 49 \neq 3^2 + 4^2 = 9 + 16 = 25$!
Here is how we multiply two binomial expressions (expressions with **2** terms) and why. Everyone agrees that $A(x+y) = Ax + Ay$. Replace A with $(x+y)$:

$$(x+y)^2 = \underbrace{(x+y)(x+y)}_{\text{This is } A(x+y)!} = \underbrace{(x+y) \cdot x}_{A \cdot x} + \underbrace{(x+y) \cdot y}_{A \cdot y} = x^2 + yx + xy + y^2 = x^2 + 2xy + y^2$$

Each "term" in the first bracket (the terms are x and y) is multiplied by each term in the second (where the terms are again x and y.) Apply that reasoning to $(a+b)(c+d)$:

$$\underbrace{F}_{\text{FIRST}} \quad \underbrace{O}_{\text{OUTSIDE}} \quad \underbrace{I}_{\text{INSIDE}} \quad \underbrace{L}_{\text{LAST}} : (a+b)(c+d) = \underbrace{ac}_{\text{FIRST TERMS}} + \underbrace{ad}_{\text{OUTSIDE TERMS}} + \underbrace{bc}_{\text{INSIDE TERMS}} + \underbrace{bd}_{\text{LAST TERMS}}$$

Example 1) Expand and simplify: (a) $(x+3)(x-2)$ (b) $(2a-3b)(2a+3b)$

Solution (a) $(x+3)(x-2) = \underbrace{x^2}_{F} - \underbrace{2x}_{O} + \underbrace{3x}_{I} - \underbrace{6}_{L} = x^2 + x - 6$

(b) $(2a-3b)(2a+3b) \quad \underset{\text{This is an example of Difference of Squares.}}{=} \quad \underbrace{4a^2}_{F} + \underbrace{6ab}_{O} - \underbrace{6ab}_{I} - \underbrace{9b^2}_{L} = 4a^2 - 9b^2$

Example 2) Expand and simplify:

(a) $(a+b+c)(x+y+z)$ (b) $(a+b+c)^2$ (c) $(a+b+c+d)^2$

Solution

(a) $(a+b+c)(x+y+z) \quad \underset{\text{Multiply each of the three terms in the first bracket with each of the three terms in the second bracket.}}{=} \quad ax + ay + az + bx + by + bz + cx + cy + cz$

(b) $(a+b+c)^2 = (a+b+c)(a+b+c)$
$= a^2 + ab + ac + ba + b^2 + bc + ca + cb + c^2$
$\underset{\text{Collect like terms.}}{=} \quad a^2 + b^2 + c^2 + 2ab + 2ac + 2cb$

(c) $(a+b+c+d)^2 \quad \underset{\text{using our experience from Example 2(b)...}}{=} \quad a^2 + b^2 + c^2 + d^2 + 2ab + 2ac + 2ad + 2bc + 2bd + 2cd$

Note: In (a), there are 3 terms in each bracket. Since each term in the first is multiplied by each term in the second, the product has $3 \times 3 = 9$ terms in all. Of course, in (b) and (c), we collect like terms. So while the answer to (c) appears to have 10 terms, there were $4 \times 4 = 16$ terms before we simplified by collecting like terms.

Adding and Subtracting Polynomial Fractions

To simplify when we add or subtract fractions, we need to get the **lowest common denominator**: $\frac{2}{3}+\frac{5}{9}-\frac{5}{12}=\frac{24}{36}+\frac{20}{36}-\frac{15}{36}=\frac{29}{36}$.

We use exactly the same method when adding and/or subtracting fractions with polynomials in the top and bottom.

Example 1) Simplify $\frac{3}{2a}-\frac{6}{5a}+\frac{3}{10a}$.

Solution $\frac{3}{2a}-\frac{6}{5a}+\frac{3}{10a}$

Get the **lowest** common denominator!
$= \frac{15}{10a}-\frac{12}{10a}+\frac{3}{10a}$

Simplify the numerator.
$= \frac{6}{10a}$

Reduce more if you can!
$= \frac{3}{5a}$

Example 2) Simplify $\frac{2x+1}{x-1}-\frac{x+1}{x+2}-\frac{5x+4}{x^2+x-2}$.

Solution $\frac{2x+1}{x-1}-\frac{x+1}{x+2}-\frac{5x+4}{x^2+x-2}$

Factor the denominators!
$= \frac{2x+1}{x-1}-\frac{x+1}{x+2}-\frac{5x+4}{(x-1)(x+2)}$

Get the **lowest** common denominator.
$= \frac{(2x+1)(x+2)}{(x-1)(x+2)}-\frac{(x+1)(x-1)}{(x-1)(x+2)}-\frac{5x+4}{(x-1)(x+2)}$

Expand the numerator.
$= \frac{2x^2+5x+2-(x^2-1)-(5x+4)}{(x-1)(x+2)}$

Now simplify the numerator.
$= \frac{x^2-1}{(x-1)(x+2)}$

Check for and divide out any further common factors.
$= \frac{\cancel{(x-1)}(x+1)}{\cancel{(x-1)}(x+2)} = \frac{x+1}{x+2}$

Multiplying and Dividing Polynomial Fractions

What we do with fractions having polynomials in the top and bottom is **exactly** what we do with fractions having numbers in the top and bottom.

$$\frac{4 \cdot 7^3}{2 \cdot 5^5} \times \frac{2^4 \cdot 5^4}{4^2 \cdot 7 \cdot 11} \quad \boxed{\text{Get common factors and bases.}} \quad = \frac{2^2 \cdot 7^3 \cdot 2^4 \cdot 5^4}{2 \cdot 5^5 \cdot 2^4 \cdot 7 \cdot 11} \quad \boxed{\text{Divide out the common factors.}} \quad = \frac{2 \cdot 7^2}{5 \cdot 11} \quad \boxed{\text{In the numerical case, work out the final value.}} \quad = \frac{98}{55} \text{ or } = 1\frac{43}{55}$$

Let's do the same question, setting it up so that it starts as **DIVISION!**

$$\frac{4 \cdot 7^3}{2 \cdot 5^5} \div \frac{4^2 \cdot 7 \cdot 11}{2^4 \cdot 5^4} \quad \boxed{\text{Invert and multiply!}} \quad = \frac{4 \cdot 7^3}{2 \cdot 5^5} \times \frac{2^4 \cdot 5^4}{4^2 \cdot 7 \cdot 11} \quad \boxed{\text{Now repeat the steps above.}} \quad = \dots = \frac{98}{55}$$

Example 1) Simplify the rational expression $\dfrac{(x^2 - 9)}{(x-3)^3} \times \dfrac{(x^2 - 3x)^2}{x^3 + 27}$.

Solution $\dfrac{(x^2 - 9)}{(x-3)^3} \times \dfrac{(x^2 - 3x)^2}{x^3 + 27}$

$\boxed{\text{Factor first! Then divide out the common factors.}}$
$= \dfrac{(\cancel{x-3})(\cancel{x+3})(x^2)(\cancel{x-3})^2}{(\cancel{x-3})^3(\cancel{x+3})(x^2 - 3x + 9)}$

$= \dfrac{x^2}{x^2 - 3x + 9}$

Example 2) Simplify: $\dfrac{m^2 + 3mn + 2n^2}{m^2 + 2mn + n^2} \div \dfrac{m^2 + 2mn}{m^2 + mn}$

Solution $\dfrac{m^2 + 3mn + 2n^2}{m^2 + 2mn + n^2} \div \dfrac{m^2 + 2mn}{m^2 + mn}$

$\boxed{\text{Invert and multiply!}}$
$= \dfrac{m^2 + 3mn + 2n^2}{m^2 + 2mn + n^2} \times \dfrac{m^2 + mn}{m^2 + 2mn}$

$\boxed{\text{Factor and divide out common factors!}}$
$= \dfrac{(\cancel{m+n})(\cancel{m+2n})}{(\cancel{m+n})^2} \times \dfrac{\cancel{m}(\cancel{m+n})}{\cancel{m}(\cancel{m+2n})}$

$= 1$

Polynomial Division

Keep this example in mind: $3 \overline{)28}$ gives quotient 9 and -27, remainder 1. | 3 divides into 28 approximately 9 times. Multiply 9 by 3 and subtract to get the remainder.

$D \equiv$ DIVIDEND $\quad Q \equiv$ QUOTIENT $\quad R \equiv$ REMAINDER $\quad d \equiv$ DIVISOR

$$\frac{D}{d} = Q + \frac{R}{d} \quad \text{(which, in the example above, gives...)} \quad \Rightarrow \quad \frac{28}{3} = 9 + \frac{1}{3} \qquad D = Q \cdot d + R \quad \text{(which, in the example above, gives...)} \quad \Rightarrow \quad 28 = 9 \cdot 3 + 1$$

Example 1) Divide the polyonomial $x^3 - 2x^2 + 5x - 7$ by $x+2$.

Solution $x+2$ divides into $x^3 - 2x^2 + 5x - 7$ approximately x^2 times, just like "3 into 28" above. Multiply $x+2$ by x^2 and subtract to get the remainder. And so on...

$$\begin{array}{r} x^2 - 4x + 13 \\ x+2 \overline{) x^3 - 2x^2 + 5x - 7} \\ -(x^3 + 2x^2) \\ \hline -4x^2 + 5x - 7 \\ -(-4x^2 - 8x) \\ \hline 13x - 7 \\ -(13x + 26) \\ \hline -33 \end{array}$$

Therefore, $\dfrac{x^3 - 2x^2 + 5x - 7}{x+2} \stackrel{D/d = Q + R/d}{=} x^2 - 4x + 13 - \dfrac{33}{x+2}$

Also, $x^3 - 2x^2 + 5x - 7 \stackrel{D = Q \cdot d + R}{=} (x^2 - 4x + 13)(x+2) - 33$

Example 2) Divide the polyonomial $x^3 + 5x - 1$ by $x-1$.

Solution For convenience, let's write $x^3 + 5x - 1 = x^3 + 0x^2 + 5x - 1$.

$$\begin{array}{r} x^2 + x + 6 \\ x-1 \overline{) x^3 + 0x^2 + 5x - 1} \\ -(x^3 - x^2) \\ \hline x^2 + 5x - 1 \\ -(x^2 - x) \\ \hline 6x - 1 \\ -(6x - 6) \\ \hline 5 \end{array}$$

Therefore, $\dfrac{x^3 + 5x - 1}{x-1} \stackrel{D/d = Q + R/d}{=} x^2 + x + 6 + \dfrac{5}{x-1}$

Also, $x^3 + 5x - 1 \stackrel{D = Q \cdot d + R}{=} (x^2 + x + 6)(x-1) + 5$

Two For You – Multiplying Expressions—FOIL

Expand and simplify each of the following expressions:

1) $(x+3y)(2x-5y+1)$ 2) $(a+2b-3c)^2$

Answers   1) $2x^2+xy+x-15y^2+3y$ 2) $a^2+4b^2+9c^2+4ab-6ac-12bc$

Two For You – Adding and Subtracting Polynomial Fractions

Simplify each of the following rational expressions:

1) $\dfrac{2x}{x-5} - \dfrac{x}{x-3} + \dfrac{1}{x^2-8x+15}$ 2) $\dfrac{x-1}{x^2-16} + \dfrac{x}{x^2-5x+4} - \dfrac{1}{x^2+3x-4}$

Answers   1) $\dfrac{x^2-x+1}{(x-5)(x-3)}$ 2) $\dfrac{2x^2+x+5}{(x-4)(x+4)(x-1)}$

Two For You – Multiplying and Dividing Polynomial Fractions

Simplify each of the following rational expressions:

1) $\dfrac{(x^2+5x+6)^2}{(x^2+6x+9)(x^2-9)} \times \dfrac{x^3-27}{x^2+4x+4}$

2) $\dfrac{(a^3+5a^2)^3}{a^2+10a+25} \div \dfrac{a^8(a^3+125)}{a^2-5a+25}$

Answers 1) $\dfrac{x^2+3x+9}{x+3}$ 2) $\dfrac{1}{a^2}$

Two For You – Polynomial Division

Divide the polynomial by the linear factor. Write your answer in the form $D = Q \cdot d + R$.

1) x^3+2x^2+3x+4 by $x-1$ 2) x^4-x^3-5x+4 by $x-2$

Answers 1) $x^3+2x^2+3x+4 = (x^2+3x+6)(x-1)+10$

2) $x^4-x^3-5x+4 = (x^3+x^2+2x-1)(x-2)+2$

Partial Fractions: Preliminaries

$\dfrac{1}{15} = \dfrac{1}{5 \cdot 3} = \dfrac{2}{5} - \dfrac{1}{3}$ Take a fraction whose denominator consists of **two or more** factors and rewrite it as a sum or difference of fractions, each of whose bottoms is **one** of the factors. Why bother? Well, in calculus we will want to compute this integral:

$\displaystyle\int \dfrac{x^2+3}{x^3-x}\,dx \;\overset{\text{Factor the bottom.}}{=}\; \int \dfrac{x^2+3}{(x-1)x(x+1)}\,dx$. On first glance, this does not look easy.

Using "partial fractions" we can write $\dfrac{x^2+3}{(x-1)x(x+1)} = \dfrac{2}{x-1} - \dfrac{3}{x} + \dfrac{2}{x+1}$. How?

By using the methods in the next two sections! Since $\displaystyle\int \dfrac{1}{x \pm a}\,dx = \ln|x \pm a| + C,$ the integral of the right side is very easy! So, we want to take $\dfrac{\text{polynomial}}{\text{another polynomial}}$, factor the bottom, and use partial fractions to find simpler polynomial fractions. Here are the rules.

(1) The degree of the numerator **must be less** than the degree of the denominator. So, for example, we use polynomial division (see page 24) to write $\dfrac{x^2+1}{x^2-1} = 1 + \dfrac{2}{x^2-1}$.

(2) By getting a common denominator on the right, convince yourself that
$$\dfrac{2x^2+3}{(x-1)^3} = \dfrac{5}{(x-1)^3} + \dfrac{4}{(x-1)^2} + \dfrac{2}{x-1}.$$
This is how we deal with repeated linear factors in the bottom.

To summarize, an expression such as $\dfrac{ax+b}{(x+c)^n}$ can always be rewritten in the form
$$\dfrac{ax+b}{(x+c)^n} = \dfrac{A_n}{(x+c)^n} + \dfrac{A_{n-1}}{(x+c)^{n-1}} + \ldots + \dfrac{A_1}{x+c}.$$

(3) If one factor in the denominator is an "irreducible" quadratic (a quadratic with non-real roots) such as x^2+1, then we use a fraction of the form $\dfrac{Ax+B}{x^2+1}$. The rule for repeated linear factors in (2) works similarly for repeated irreducible quadratics.

(4) What about irreducible cubics or quartics or polynomials of even higher degree? Big Math Theorem: **THERE ARE NONE!** Every polynomial factors into linear and irreducible quadratic terms.

Example 1) Setup $\dfrac{x+1}{(x^2+1)^2(x-3)^3}$ in terms of partial fractions.

Solution $\dfrac{x+1}{(x^2+1)^2(x-3)^3} = \dfrac{Ax+B}{(x^2+1)^2} + \dfrac{Cx+D}{x^2+1} + \dfrac{E}{(x-3)^3} + \dfrac{F}{(x-3)^2} + \dfrac{G}{x-3}$

Partial Fractions: Distinct Linear Factors

Example 1) Write $\dfrac{x+1}{x^2-5x+4} = \dfrac{x+1}{(x-4)(x-1)}$ in terms of partial fractions.

Solution Set $\dfrac{x+1}{(x-4)(x-1)} = \dfrac{A}{x-4} + \dfrac{B}{x-1} \overset{\text{common denominator}}{=} \dfrac{A(x-1)+B(x-4)}{(x-4)(x-1)} = \dfrac{(A+B)x-A-4B}{(x-4)(x-1)}$.

We need to find the values for A and B that make this work. Since the denominators are the same, we compare the numerators.

Coefficient Method: We need $x+1 \overset{\text{These expressions must be IDENTICAL!}}{=} (A+B)x - A - 4B$. Compare coefficients.

x coefficient: $1 = A+B$ (The coefficient of x is 1; the coefficient of $(A+B)x$ is $A+B$.)

constant: $\quad 1 = -A - 4B$ (The constant on the left is 1; the constant on the right is $-A-4B$.)

Add the equations: $2 = -3B$ and so $B = -\dfrac{2}{3}$. Substituting into either equation gives $A = \dfrac{5}{3}$.

Substitution Method: Look at the equation $x+1 \overset{\text{These expressions must be EQUAL for every } x \in \mathbb{R}!}{=} A(x-1) + B(x-4)$.

Note that $x=1$ makes $x-1=0$ and so it is easy to find B. Similarly, $x=4$ makes A easy to find. We need $x+1 = A(x-1) + B(x-4)$.

Let $x=1$: $2 = A(0) + B(-3)$ and so $B = -\dfrac{2}{3}$. Let $x=4$: $5 = A(3) + B(0) = 3A$ and so $A = \dfrac{5}{3}$.

We find the same answers for A and B as we did using the coefficient method. It would be distressing if this were not the case! So, we have shown $\dfrac{x+1}{(x-4)(x-1)} = \dfrac{\frac{5}{3}}{x-4} - \dfrac{\frac{2}{3}}{x-1}$.

Example 2) Write $\dfrac{x^2+1}{x^2+x-6} = \dfrac{x^2+1}{(x+3)(x-2)}$ in terms of partial fractions.

Solution Whenever the top has the same or larger degree (the highest power of x) as the bottom, **divide** first. Using polynomial division, $\dfrac{x^2+1}{x^2+x-6} = 1 + \dfrac{-x+7}{x^2+x-6} = 1 + \dfrac{-x+7}{(x+3)(x-2)}$.

$\dfrac{-x+7}{(x+3)(x-2)} = \dfrac{A}{x+3} + \dfrac{B}{x-2} = \dfrac{A(x-2)+B(x+3)}{(x+3)(x-2)}$.

Use the substitution method to solve $-x+7 = A(x-2) + B(x+3)$.

$x=2$: $5 = 5B$ and so $B=1$; $\quad x=-3$: $10 = -5A$ and so $A = -2$.

So, we have $\dfrac{x^2+1}{x^2+x-6} = 1 + \dfrac{-2}{x+3} + \dfrac{1}{x-2}$.

Partial Fractions: Repeated Linear Factors

The method for partial fractions, to be properly explained, really needs tools from linear algebra. No time! No space! So I hope the recipe below is, at least, easy to follow.

Example 1) Write $\dfrac{x^2+2x+1}{(x-1)^2(x+3)}$ in terms of partial fractions.

Solution Set $\dfrac{x^2+2x+1}{(x-1)^2(x+3)} = \dfrac{A}{(x-1)^2} + \dfrac{B}{x-1} + \dfrac{C}{x+3}$

$\overset{\text{common denominator}}{=} \dfrac{A(x+3) + B(x+3)(x-1) + C(x-1)^2}{(x-1)^2(x+3)}$

We need to find the values for A, B, and C so that

$x^2+2x+1 = A(x+3) + B(x+3)(x-1) + C(x-1)^2$. Let's first use substitution.

$x=1$: $4=4A$ and so $A=1$.

$x=-3$: $4=16C$ and so $C=\dfrac{1}{4}$.

We have run out of **easy** numbers to substitute for x.* Let's compare the coefficient of x^2:

On the left, we have $x^2 = 1x^2$, so the coefficient is 1.

On the right, if we expand, we will get $Bx^2 + Cx^2 = (B+C)x^2$.

For these expresssions to be **identical**, we have $1 = B+C$, that is, $1 = B + \dfrac{1}{4}$ and so $B = \dfrac{3}{4}$.

Therefore, $\dfrac{x^2+2x+1}{(x-1)^2(x+3)} = \dfrac{1}{(x-1)^2} + \dfrac{\frac{3}{4}}{x-1} + \dfrac{\frac{1}{4}}{x+3}$.

Example 2) Set up $\dfrac{x^5-1}{x^3(x+1)^2(x+3)}$ in terms of partial fractions.

Do not solve for the values of A, B, C, etc.

Solution Set $\dfrac{x^5-1}{x^3(x+1)^2(x+3)} = \dfrac{A}{x^3} + \dfrac{B}{x^2} + \dfrac{C}{x} + \dfrac{D}{(x+1)^2} + \dfrac{E}{x+1} + \dfrac{F}{x+3}$

*In fact, we could substitute x equal to **any** number other than 1 and -3. We will get an equation in A, B, and C which will yield the correct value for B. It is just easier resorting to the coefficient method when we have used up the easy x values to plug in.

Partial Fractions: Irreducible Quadratics

When we want to use partial fractions with a quadratic whose roots are not real numbers*, we allow for both an x term plus a constant in the numerator.**

Example 1) Write $\dfrac{x+2}{(x^2+1)(x+1)}$ in terms of partial fractions.

Solution Set $\dfrac{x+2}{(x^2+1)(x+1)} = \dfrac{Ax+B}{x^2+1} + \dfrac{C}{x+1} = \dfrac{(Ax+B)(x+1)+C(x^2+1)}{(x^2+1)(x+1)}$.

We need to find the values for A, B, and C so that
$x+2 = (Ax+B)(x+1) + C(x^2+1)$.

$x = -1$: $1 = 2C$ and so $C = \dfrac{1}{2}$.

x^2 coefficient: $0 = A + C$ and so $A = -C = -\dfrac{1}{2}$.

constant: $2 = B + C$ and so $B = 2 - C = \dfrac{3}{2}$.

Therefore, $\dfrac{x+2}{(x^2+1)(x+1)} = \dfrac{-\frac{1}{2}x+\frac{3}{2}}{x^2+1} + \dfrac{\frac{1}{2}}{x+1} = \dfrac{1}{2}\left(\dfrac{-x+3}{x^2+1} + \dfrac{1}{x+1}\right)$.

Example 2) Set up $\dfrac{x^5-1}{(x^2+x+2)^3(x^2+1)(x+3)^3 x}$ in terms of partial fractions. Do not solve for the values of A, B, C, etc.

Solution Set $\dfrac{x^5-1}{(x^2+x+2)^3(x^2+1)(x+3)^3 x}$
$= \dfrac{Ax+B}{(x^2+x+2)^3} + \dfrac{Cx+D}{(x^2+x+2)^2} + \dfrac{Ex+F}{x^2+x+2} + \dfrac{Gx+H}{x^2+1} + \dfrac{I}{(x+3)^3} + \dfrac{J}{(x+3)^2} + \dfrac{K}{x+3} + \dfrac{L}{x}$.

*A reminder of how to quickly check this: if $ax^2+bx+c=0$ has non-real roots, the discriminant b^2-4ac will be negative.

**What about irreducible cubics or quartics or polynomials of higher degree? No need! *Big Theorem: Every polynomial factors as a product of linear terms and irreducible quadratics!*

Two For You – Partial Fractions: Preliminaries

1) Setup $\dfrac{x^2+3x-7}{(2x+1)^3(x-3)^2 x}$ in terms of partial fractions.

2) Setup $\dfrac{x^2-7}{(x^2+x+1)^3(x^2+7)^2 x^2(x-1)}$ in terms of partial fractions.

Answers

1) $\dfrac{x^2+3x-7}{(2x+1)^3(x-3)^2 x} = \dfrac{A}{(2x+1)^3} + \dfrac{B}{(2x+1)^2} + \dfrac{C}{2x+1} + \dfrac{D}{(x-3)^2} + \dfrac{E}{x-3} + \dfrac{F}{x}$

2) $\dfrac{x^2-7}{(x^2+x+1)^3(x^2+7)^2 x^2(x-1)} =$

$\dfrac{Ax+B}{(x^2+x+1)^3} + \dfrac{Cx+D}{(x^2+x+1)^2} + \dfrac{Ex+F}{x^2+x+1} + \dfrac{Gx+H}{(x^2+7)^2} + \dfrac{Ix+J}{x^2+7} + \dfrac{K}{x^2} + \dfrac{L}{x} + \dfrac{M}{x-1}$

Two For You – Partial Fractions: Distinct Linear Factors

1) Write $\dfrac{x+5}{x^3-x}$ in terms of partial fractions.

2) Write $\dfrac{x^3}{(x+2)(x-2)}$ in terms of partial fractions.

Answers 1) $\dfrac{3}{x-1} - \dfrac{5}{x} + \dfrac{2}{x+1}$ 2) $x + \dfrac{2}{x-2} + \dfrac{2}{x+2}$

Two For You – Partial Fractions: Repeated Linear Factors

1) Write $\dfrac{2x-1}{(x+1)^2(x-1)}$ in terms of partial fractions.

2) Write $\dfrac{x-2}{x^3(x+2)}$ in terms of partial fractions.

Answers 1) $\dfrac{\frac{3}{2}}{(x+1)^2} - \dfrac{\frac{1}{4}}{x+1} + \dfrac{\frac{1}{4}}{x-1}$ $\boxed{\text{You may prefer ...}}$ $= \dfrac{3}{2(x+1)^2} - \dfrac{1}{4(x+1)} + \dfrac{1}{4(x-1)}$

2) $-\dfrac{1}{x^3} + \dfrac{1}{x^2} - \dfrac{1}{2x} + \dfrac{1}{2(x+2)}$

Two For You – Partial Fractions: Irreducible Quadratics

1) Write $\dfrac{x+2}{(x^2+x+1)(x-1)}$ in terms of partial fractions.

2) Write $\dfrac{x-3}{(x^2+1)x^2}$ in terms of partial fractions.

Answers 1) $\dfrac{-x-1}{x^2+x+1} + \dfrac{1}{x-1}$ 2) $\dfrac{-x+3}{x^2+1} - \dfrac{3}{x^2} + \dfrac{1}{x}$

Finding the Equation of a Line

Given the slope and a point, or two points, there are **lots** of ways to find the equation of a line. The easiest method to cover all scenarios uses the equation $y - y_1 = m(x - x_1)$, where m is the slope and (x_1, y_1) is a point on the line.

Example 1) Find the equation of the line with slope -3 through the point $(2, -5)$.

Solution $\boxed{x_1 = 2, y_1 = -5, m = -3}$
$$y - (-5) = -3(x - 2)$$
$$y + 5 = -3x + 6$$
$$y = -3x + 1$$

Example 2) Find the equation of the line passing through the points $(2, 7)$ and $(4, 12)$.

Solution The slope $m = \dfrac{y_2 - y_1}{x_2 - x_1} = \dfrac{12 - 7}{4 - 2} = \dfrac{5}{2}$.

Note: $\dfrac{y_1 - y_2}{x_1 - x_2} = \dfrac{7 - 12}{2 - 4} = \dfrac{5}{2}$ as well.

Using $(2, 7)$ as (x_1, y_1):   OR   Using $(4, 12)$ as (x_1, y_1):

$y - 7 = \dfrac{5}{2}(x - 2)$ $\qquad\qquad$ $y - 12 = \dfrac{5}{2}(x - 4)$

$y - 7 = \dfrac{5}{2}x - 5$ $\qquad\qquad$ $y - 12 = \dfrac{5}{2}x - 10$

$y = \dfrac{5}{2}x + 2$ $\qquad\qquad\qquad$ $y = \dfrac{5}{2}x + 2$

Example 3) Find the equation of the line passing through the points $(2, 1)$ and $(4, 1)$.

Solution The slope $m = \dfrac{y_2 - y_1}{x_2 - x_1} = \dfrac{1 - 1}{4 - 2} = \dfrac{0}{2} = 0$.

Using $(2, 1)$ as (x_1, y_1):   OR   Using $(4, 1)$ as (x_1, y_1):

$y - 1 = 0(x - 2)$ $\qquad\qquad$ $y - 1 = 0(x - 4)$

$y - 1 = 0$ $\qquad\qquad\qquad\quad$ $y - 1 = 0$

$\quad y = 1$ $\qquad\qquad\qquad\qquad\quad$ $y = 1$

Slope m and y Intercept b

In the equation of the line $y = mx + b$, m is the slope and b is the y intercept.

Example 1) Find the slope and the y intercept for each of the following lines.

a) $y = 3x + 5$ b) $y = 7 - 2x$ c) $y = 4x$ d) $y = -7$ e) $2x + 3y = 4$

f) $4y - 5x - 2 = 0$ g) $x = 4$ h) $y = 0$ i) $x = 0$ j) $\dfrac{x}{3} + \dfrac{y}{2} = 1$

Solution

a) $m = 3, b = 5$ b) $m = -2, b = 7$ c) $m = 4, b = 0$ d) $m = 0, b = -7$

e) Rewrite the equation in the form $y = mx + b$.

$2x + 3y = 4 \Leftrightarrow 3y = 4 - 2x \Leftrightarrow y = -\dfrac{2}{3}x + \dfrac{4}{3}$ and so $m = -\dfrac{2}{3}$ and $b = \dfrac{4}{3}$

f) Rewrite the equation in the form $y = mx + b$.

$4y - 5x - 2 = 0 \Leftrightarrow 4y = 5x + 2 \Leftrightarrow y = \dfrac{5}{4}x + \dfrac{1}{2}$ and so $m = \dfrac{5}{4}$ and $b = \dfrac{1}{2}$

g) The slope is undefined (or infinite). This is the vertical line where x **always equals 4** while y can be any real number. It is parallel to the y axis and **there is no y intercept**.

h) $m = b = 0$. ($y = 0$ is the equation of the x axis.)

i) The slope is undefined (or infinite). $x = 0$ is the equation of the y axis, so there are **LOTS** of y intercepts!

j) Rewrite the equation in the form $y = mx + b$.

$\dfrac{x}{3} + \dfrac{y}{2} = 1 \Leftrightarrow \dfrac{y}{2} = 1 - \dfrac{x}{3} \Leftrightarrow y = -\dfrac{2}{3}x + 2$ and so $m = -\dfrac{2}{3}$ and $b = 2$

Graphing a Straight Line Using $y = mx + b$

When asked to graph a straight line, many of you will jump right to a table of values. There is a faster method that doesn't involve any rough work.

Example 1) Graph $y = -\dfrac{2}{3}x + 4$.

Solution
Step 1: The y intercept is 4. Plot $(0, 4)$ on the graph.

Step 2: Use the slope to get a pattern of how to move from one point to the next.
$m = \dfrac{rise}{run} = -\dfrac{2}{3} = \dfrac{-2}{3} \begin{array}{l} \to \text{down } 2 \\ \to \text{right } 3 \end{array}$

(We could have used $\dfrac{2}{-3}$ and gone up 2 and left 3.)

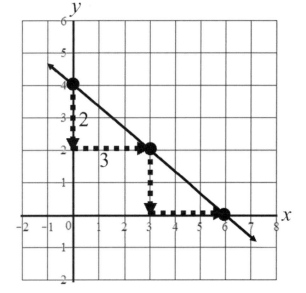

Step 3: Starting from your y intercept, use the pattern in step 2 to add a few more points to the graph.

Step 4: Draw a line through your points.

However, what if the next point after the y intercept jumps off of the grid you were given? No need to make a whole new grid; just a bit of "backwards" thinking is required!

Example 2) Graph $y = 2x + 6$.

Solution This time, when we start at the y intercept and use the slope to find a pattern to move from one point to the next,
$m = \dfrac{rise}{run} = \dfrac{2}{1} \begin{array}{l} \to \text{up } 2 \\ \to \text{right } 1 \end{array}$
we get to the point $(1, 8)$, which is off of our grid.

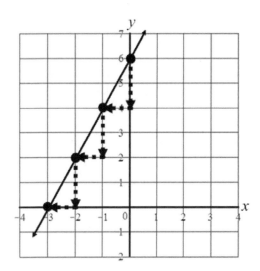

Instead of thinking how to get to the next point, think of how to get to the point **before** the point you are at. This means instead of counting up 2, right 1, we should count **down** 2 and **left** 1.

Distance between Two Points and Distance from a Point to a Line or Plane

Example 1) Find the distance from $(-1, 5)$ to $(6, 3)$.

Solution Distance $= \overset{\sqrt{(x_2-x_1)^2+(y_2-y_1)^2}}{=} \sqrt{(6-(-1))^2+(3-5)^2} = \sqrt{7^2+(-2)^2} = \sqrt{53}$

Example 2) Find the (perpendicular) distance from the point $(3, 4)$ to the line $4x - 5y = 7$.

Solution The distance from (x_0, y_0) to $Ax + By + C = 0$ is given by the formula $\frac{|Ax_0 + By_0 + C|}{\sqrt{A^2 + B^2}}$. So, from the point $(3, 4)$ to the line $4x - 5y = 7$,

distance $\overset{A=4, B=-5, \text{ and } C=-7}{=} \frac{|(4)(3) + (-5)(4) - 7|}{\sqrt{4^2 + 5^2}} = \frac{|-15|}{\sqrt{41}} = \frac{15}{\sqrt{41}}$.

Example 3) Find the distance from $(3, 4, -1)$ to $(0, 5, -2)$.

Solution distance $= \sqrt{(x_2-x_1)^2 + (y_2-y_1)^2 + (z_2-z_1)^2}$
$= \sqrt{(0-3)^2 + (5-4)^2 + (-2-(-1))^2} = \sqrt{(-3)^2 + (1)^2 + (-1)^2} = \sqrt{11}$

Example 4) Find the perpendicular distance from the point $(1, 2, 3)$ to the plane $2x + 4y - 5z = 7$.

Solution The distance from (x_0, y_0, z_0) to $Ax + By + Cz + D = 0$ is given by the formula $\frac{|Ax_0 + By_0 + Cz_0 + D|}{\sqrt{A^2 + B^2 + C^2}}$. So, from the point $(1, 2, 3)$ to the plane $2x + 4y - 5z = 7$,

distance $\overset{A=2, B=4, C=-5, \text{ and } D=-7}{=} \frac{|(2)(1) + (4)(2) + (-5)(3) - 7|}{\sqrt{2^2 + 4^2 + (-5)^2}}$

$= \frac{|-12|}{\sqrt{45}} = \frac{12}{\sqrt{45}} = \frac{12}{3\sqrt{5}} = \frac{4}{\sqrt{5}} \overset{\text{for fans of BOB (Back Of the Book!)}}{=} \frac{4\sqrt{5}}{5}$.

Two For You – Finding the Equation of a Line

1) Find the equation of the line through $(-1, -5)$ and $(3, 7)$.
2) Find the equation of the line with x intercept 4 and y intercept 7.
(Hint: use the points $(4, 0)$ and $(0, 7)$.)

Answers 1) $y = 3x - 2$ 2) $y = -\dfrac{7}{4}x + 7$

Two For You – Slope m and y Intercept b

Find the slope and the y intercept for the following lines:
1) $\pi y - 2x = 1$ 2) $3x - 4 = 0$

Answers 1) $m = \dfrac{2}{\pi}$, $b = \dfrac{1}{\pi}$ 2) undefined slope, no y intercept

Two For You – Graphing a Straight Line Using $y = mx + b$

Graph each line using the grid provided.

1) $y = \dfrac{-1}{2}x - 1$

2) $y = -3x - 5$

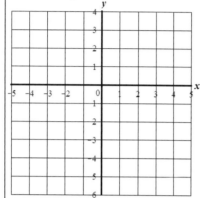
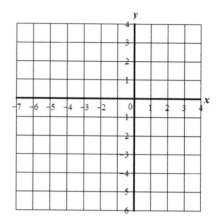

Answers 1) y int $= -1$; $m = -1/2$; use pattern down 1, right 2.
2) y int $= -5$; $m = -3/1$; use pattern up 3, left 1.

Two For You – Distance between a Point and a Point/Line/Plane

1) Find the distance between $(-3, 0, 1)$ and $(5, 2, 0)$.

2) Find the distance from the point $(1, 2, 3)$ to the plane $3x - 7z = 0$.
(Hint: $A = 3$, $B = 0$, $C = -7$, and $D = 0$.)

Answers 1) $\sqrt{69}$ 2) $\dfrac{18}{\sqrt{58}} = \dfrac{9\sqrt{58}}{29}$

Visually Identifying Slopes of Lines

This page is all about visually recognizing the slope of a line. Assume the scales on the x and y axes are equal in all cases. Slope tells you how much y increases for a unit change in x. So a slope of 3 means if x **increases** by 1, then y **increases** by 3. If x goes up by 5, then y goes up by $3 \times 5 = 15$, that is, y goes up 3 times as much as x. A slope of -3 means that if x **increases** by 1, then y **decreases** by 3.

Given points (x_1, y_1) and (x_2, y_2), the slope of the segment joining these points is

$$m = \frac{y_2 - y_1}{x_2 - x_1} = \frac{y_1 - y_2}{x_1 - x_2} = \frac{\Delta y}{\Delta x} = \frac{\text{rise}}{\text{run}} = \frac{\text{change in } y}{\text{change in } x}$$

Warning! Don't confuse the symbol "m" for slope with the lines labelled "m" below.

Positive Slope
y **goes up**
as x **goes up**.

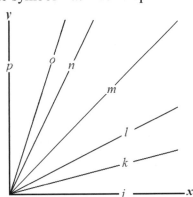

Slope p: undefined
Slope $o = 4$
Slope $n = 2$
Slope $m = 1$
Slope $l = 0.5$
Slope $k = 0.25$
Slope $j = 0$

Negative slope
y **goes down**
as x **goes up**.

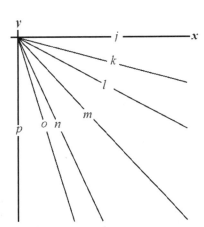

Slope $j = 0$
Slope $k = -0.25$
Slope $l = -0.5$
Slope $m = -1$
Slope $n = -2$
Slope $o = -4$
Slope p: undefined

Note:

Horizontal lines (line j in both pictures) have **slope 0**.

Vertical lines (line p in both pictures) have **undefined (or infinite) slope**.

Parallel and Perpendicular Lines

If line l_1 with slope m_1 is parallel to line l_2 with slope m_2, then $m_1 = m_2$.

If line l_1 with slope $m_1 \neq 0$ is perpendicular to line l_2 with slope m_2, then $m_2 = -\dfrac{1}{m_1}$, that is, the slope of l_2 is the **negative reciprocal** of the slope of l_1.

In the case of perpendicular lines, if $m_1 = 0$, then l_1 is horizontal, that is, l_1 is parallel to the x axis. In this case, l_2 is vertical, that is, parallel to the y axis, and its slope is undefined, that is, l_2 has "infinite slope".

Example 1) Find the equation of the line which is (a) parallel (b) perpendicular to the line $y = -3x + 5$ which passes through the point $(-4, 3)$.

Solution (a) The slope of a parallel line is -3. Using $y - y_1 = m(x - x_1)$, we have $y - 3 = -3(x - (-4))$ and so $y = -3x - 9$.

(b) The slope of a perpendicular line is $-\dfrac{1}{(-3)} = \dfrac{1}{3}$. Therefore,

$y - 3 = \dfrac{1}{3}(x + 4)$ and so $y = \dfrac{1}{3}x + \dfrac{13}{3}$.

Example 2) Find the equation of the line which is (a) parallel (b) perpendicular to the line $y = -1$ which passes through the point $(1, 2)$.

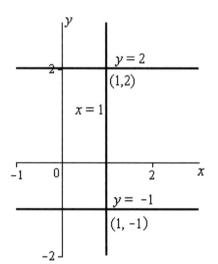

Solution (a) The slope of a parallel line is 0 and y is always 2: $y = 2$ [y=mx+b, m=0 and b=2]

(b) The slope of a perpendicular line is undefined and on this line, x is always 1: $x = 1$

Finding Tangent and Normal Lines to a Curve

Let $y = f(x)$. The slope of the tangent line to this function at the point $(a, f(a))$ is given by $f'(a)$ and the slope of the normal line by $-\dfrac{1}{f'(a)}$.

Example 1) Let $f(x) = x^3 - 8x + 9$.
Find the equation of the tangent
and normal lines at the point where $x = 2$.

Solution $f'(x) = 3x^2 - 8$.
Now $f(2) = 1$ and $f'(2) = 4$. Using
$y - y_1 = m(x - x_1)$, the tangent line
is $y - 1 = 4(x - 2)$ and so $y = 4x - 7$.
For the normal line, we still use the
point $(2, 1)$ but the slope is $-\dfrac{1}{4}$.
Therefore, the normal equation is
$y - 1 = -\dfrac{1}{4}(x - 2)$ and so
$y = -\dfrac{1}{4}x + \dfrac{1}{2} + 1 = -\dfrac{1}{4}x + \dfrac{3}{2}$.

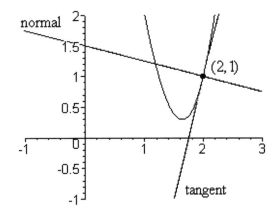

Example 2) Let $xy + e^{y-1} = 2$. Find the equation of the tangent and normal lines at the point where $y = 1$.

Solution Substituting $y = 1$ gives (remember $e^0 = 1$) $x + 1 = 2$ and so $x = 1$.
Differentiating implicitly: $x\dfrac{dy}{dx} + y + e^{y-1}\dfrac{dy}{dx} = 0$ ∴ $\dfrac{dy}{dx}(x + e^{y-1}) = -y$
and so $\dfrac{dy}{dx} = \dfrac{-y}{x + e^{y-1}}$.
At $(1,1)$, $\dfrac{dy}{dx} = -\dfrac{1}{2}$.
The tangent line is $y - 1 = -\dfrac{1}{2}(x - 1)$ and so $y = -\dfrac{1}{2}x + \dfrac{3}{2}$.
The normal line is $y - 1 = 2(x - 1)$ and so $y = 2x - 1$.

Solving a Linear Equation

When solving for a variable, everyone knows the basics of "get x by itself". However, carrying out that plan can get a little tricky when there are variables on both sides of the equation. All you need to remember is this fundamental rule in mathematics:

What you do to one side, you do to the other!

Example 1) Solve for x.

$7x + 3 = 2x - 2$
$-2x \qquad -2x$ 	Subtract $2x$ from both sides to get all the variables on the left.
$5x + 3 = -2$
$\quad -3 \quad -3$ 	Subtract 3 from both sides to get all the number terms on the right.
$5x = -5$
$\dfrac{5x}{5} = \dfrac{-5}{5}$ 	Divide both sides by the coefficient in front of the variable.
$x = -1$

Example 2) Solve for y.

$3(y-2) - 2(y+3) = 4(y-4)$
$3y - 6 - 2y - 6 = 4y - 16$ 	Expand both sides.
$y - 12 = 4y - 16$ 	Simplify both sides.
$-4y \qquad -4y$
$-3y - 12 = -16$
$\quad +12 \quad +12$
$-3y = -4$
$y = \dfrac{4}{3}$

Example 3) Solve for x.

$\dfrac{x}{4} - \dfrac{8}{3} = \dfrac{x}{6} + \dfrac{5}{4}$

$12\left(\dfrac{x}{4} - \dfrac{8}{3}\right) = 12\left(\dfrac{x}{6} + \dfrac{5}{4}\right)$ 	Multiply both sides by the lowest common denominator.

$3x - 32 = 2x + 15$
$-2x \qquad\quad -2x$
$x - 32 = 15$
$\quad +32 \quad +32$
$x = 47$

One For You – Visually Identifying Slopes of Lines

1) Match the slopes $3, -4, 0, 0.5, -0.5,$ and undefined with the segments.

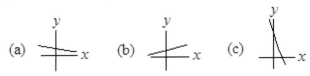

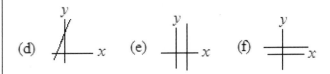

Answers (a) -0.5 (b) 0.5 (c) -4 (d) 3 (e) undefined   (f) 0

Two For You – Parallel and Perpendicular Lines

1) Find the equation of the line through the point $(1, 1)$ that is
(a) parallel to the line through the points $(1, 1)$ and $(3, 11)$
(b) perpendicular to the line through the points $(1, 1)$ and $(3, 11)$.
2) Find the equation of the line with x intercept 1 that is
(a) parallel to $y = 4$ (b) perpendicular to $y = 4$.

Answers 1)(a) $y = 5x - 4$ (b) $y = -\frac{1}{5}x + \frac{6}{5}$

2)(a) $y = 0$ (Note: here, every real number is an x intercept!)   (b) $x = 1$

Two For You – Finding Tangent and Normal Lines to a Curve

1) Find the tangent and normal lines to $y = x^2 - x$ at $x = -1$.

2) Find the tangent and normal lines to $x + y + \ln y = 4$ at $y = 1$.

Answers 1) tangent: $y = -3x - 1$ normal: $y = \dfrac{1}{3}x + \dfrac{7}{3}$

2) tangent: $y = -\dfrac{1}{2}x + \dfrac{5}{2}$ normal: $y = 2x - 5$

Two For You – Solving a Linear Equation

1) Solve for x: $4(x-4) - 2(x-3) = 3(3x+6)$

2) Solve for y: $\dfrac{3y}{4} + 2 = \dfrac{2y+3}{3}$

Answers 1) $x = -4$ 2) $y = -12$

Solving Two Linear Equations Using Substitution

In the earlier high school grades, we usually solve two linear equations by either the "substitution" or "elimination" methods. There is a method called "row reduction" which works much better when you have **three or more** equations. (See pages 46 and 47.) In fact, elimination is just a simple version of row reduction. However, for two equations, substitution is usually the best. So in this section, **substitution rules** rule!

Example 1) Solve E1: $x+3y=5$ and E2: $4x-5y=3$ using substitution.

Solution Solve for x in E1: $x+3y=5 \Rightarrow x=5-3y$. Call this equation E3.
Substitute this into E2:
$4(5-3y)-5y=3$
$\Rightarrow 20-12y-5y=3$
$\Rightarrow -17y=-17$
$\Rightarrow y=1$
Now we substitute $y=1$ into E3: $x=5-3(1)=5-3=2$

Geometrically, we have found that the two straight lines $x+3y=5$ and $4x-5y=3$ intersect at the point $(2,1)$.

Example 2) Solve E1: $5y-3x=-6$ and E2: $5x-8y=0$ using substitution.

Solution First, solve for y in E1:
$5y=3x-6 \Rightarrow 5y-3x=-6 \Rightarrow y=\dfrac{3x-6}{5}$. Call this equation E3.
Now substitute this into E2:

$5x-8\left(\dfrac{3x-6}{5}\right)=0 \quad \boxed{\text{Clear the denominator by multiplying the equation by 5.}} \Rightarrow \quad 25x-8(3x-6)=0$

$\Rightarrow 25x-24x+48=0 \Rightarrow x=-48$.

Now substitute this into the equation for E3: $y=\dfrac{3(-48)-6}{5}=\dfrac{-144-6}{5}=\dfrac{-150}{5}=-30$

Note: You can solve for x first and then find y or solve first for y and then find x. Do what seems easiest for the particular example. In Example 1, it was best to solve for x first because, well, you tell me! (Hint: can you say, "Fractions!"?)

Solving Two Linear Equations Using "Row Reduction"

Let's use row reduction to solve these pairs of equations. There are other methods, and in fact some of these are easier than row reduction when you have only two equations, **but this is the easiest method to generalize to three, four, and more linear equations.**

Example Solve each of the following pairs of linear equations:
1) $x + 3y = 5$ and $4x - 5y = 3$
2) $3x - 5y = 6$ and $5x - 8y = 0$
3) $2x - y = 7$ and $4x - 2y = 14$
4) $2x - y = 7$ and $4x - 2y = 6$

Solution 1) $\begin{bmatrix} x + 3y = 5 & E1 \\ 4x - 5y = 3 & E2 \end{bmatrix}$ $\overset{\text{Leave E1 alone!}}{\underset{-4 \times E1 + E2 = \text{New E2}}{\Leftrightarrow}}$ $\begin{bmatrix} x + 3y = 5 & E1 \\ -17y = -17 & E2 \end{bmatrix}$

From $-17y = -17$, we find $y = 1$.

Substitute $y = 1$ into the **original** E1: $x + 3(1) = 5$ and so $x = 2$. The solution is $(x, y) = (2, 1)$.

2) $\begin{bmatrix} 3x - 5y = 6 & E1 \\ 5x - 8y = 0 & E2 \end{bmatrix}$ $\overset{\frac{1}{3} \times E1 = \text{New E1}}{\underset{\text{Leave E2 alone!}}{\Leftrightarrow}}$ $\begin{bmatrix} x - \frac{5}{3}y = 2 & E1 \\ 5x - 8y = 0 & E2 \end{bmatrix}$

$\overset{\text{Leave E1 alone!}}{\underset{-5 \times E1 + E2 = \text{New E2}}{\Leftrightarrow}}$ $\begin{bmatrix} x - \frac{5}{3}y = 2 & E1 \\ \frac{1}{3}y = -10 & E2 \end{bmatrix}$

From $\frac{1}{3}y = -10$, we have $y = -30$.

Substitute $y = -30$ into the **original** E1: $3x - 5(-30) = 6$ and so $3x = -144$ and $x = -48$.

3) Solving the same way as in 1) and 2) yields the equation $0 = 0$. These lines are **COINCIDENT**, that is, both equations represent the same line. The solutions are $x \in \mathbb{R}$ and $y = 2x - 7$.

4) This time, we obtain the equation $0 = 8$. There is no solution. These are non-intersecting parallel lines.

Solving Three Linear Equations Using Row Reduction

We will use row reduction here just as we did with two equations in two unknowns on page 46. The method is **long**, tedious, and it's easy to make mechanical errors. The good news is that the steps are pretty straightforward.

Example Solve the following systems of linear equations:

1) $\begin{array}{l} x+2y-6z=5 \\ 4x+5y-21z=5 \\ -3x+3y+17z=-2 \end{array}$

2) $\begin{array}{l} 2x+y-2z=10 \\ 3x+2y+2z=1 \\ 5x+4y+3z=4 \end{array}$

Solution

1)

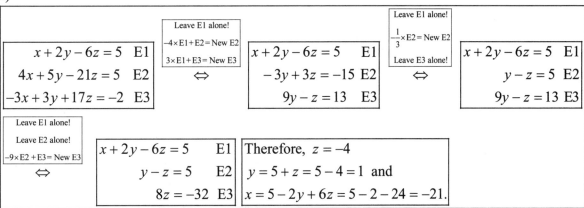

2)

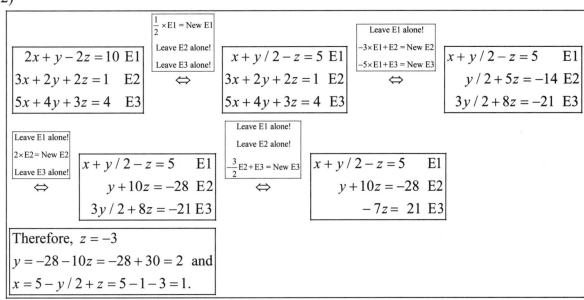

Solving Three Linear Equations REDUX: MATRICES

We are now going to, line by line, repeat the previous section using **matrices** (singular, matrix). We don't need to include $x, y,$ and z every time we write an equation. All we really **need** are the coefficients. What follows explains itself! Line by line, it follows the steps on page 47.

Example Solve the following systems of linear equations:

1) $\begin{cases} x+2y-6z=5 \\ 4x+5y-21z=5 \\ -3x+3y+17z=-2 \end{cases}$

2) $\begin{cases} 2x+y-2z=10 \\ 3x+2y+2z=1 \\ 5x+4y+3z=4 \end{cases}$

Solution

1) $\begin{pmatrix} 1 & 2 & -6 & | & 5 \\ 4 & 5 & -21 & | & 5 \\ -3 & 3 & 17 & | & -2 \end{pmatrix}$

Leave $R1 = ROW1$ alone!
$-4 \times R1 + R2 =$ New R2
$3 \times R1 + R3 =$ New R3
$\Leftrightarrow$

$\begin{pmatrix} 1 & 2 & -6 & | & 5 \\ 0 & -3 & 3 & | & -15 \\ 0 & 9 & -1 & | & 13 \end{pmatrix}$

Leave R1 alone!
$-\frac{1}{3} \times R2 =$ New R2
Leave R3 alone!
$\Leftrightarrow$

$\begin{pmatrix} 1 & 2 & -6 & | & 5 \\ 0 & 1 & -1 & | & 5 \\ 0 & 9 & -1 & | & 13 \end{pmatrix}$

Leave R1 alone!
Leave R2 alone!
$-9 \times R2 + R3 =$ New R3
$\Leftrightarrow$

$\begin{pmatrix} 1 & 2 & -6 & | & 5 \\ 0 & 1 & -1 & | & 5 \\ 0 & 0 & 8 & | & -32 \end{pmatrix}$

Therefore, $z = -4$
$y = 5 + z = 5 - 4 = 1$ and
$x = 5 - 2y + 6z = 5 - 2 - 24 = -21.$

2) $\begin{pmatrix} 2 & 1 & -2 & | & 10 \\ 3 & 2 & 2 & | & 1 \\ 5 & 4 & 3 & | & 4 \end{pmatrix}$

$\frac{1}{2} \times R1 =$ New R1
Leave R2 alone!
Leave R3 alone!
$\Leftrightarrow$

$\begin{pmatrix} 1 & .5 & -1 & | & 5 \\ 3 & 2 & 2 & | & 1 \\ 5 & 4 & 3 & | & 4 \end{pmatrix}$

Leave R1 alone!
$-3 \times R1 + R2 =$ New R2
$-5 \times R1 + R3 =$ New R3
$\Leftrightarrow$

$\begin{pmatrix} 1 & .5 & -1 & | & 5 \\ 0 & .5 & 5 & | & -14 \\ 0 & 1.5 & 8 & | & -21 \end{pmatrix}$

Leave R1 alone!
$2 \times R2 =$ New R2
Leave R3 alone!
$\Leftrightarrow$

$\begin{pmatrix} 1 & .5 & -1 & | & 5 \\ 0 & 1 & 10 & | & -28 \\ 0 & 1.5 & 8 & | & -21 \end{pmatrix}$

Leave R1 alone!
Leave R2 alone!
$-1.5 R2 + R3 =$ New R3
$\Leftrightarrow$

$\begin{pmatrix} 1 & .5 & -1 & | & 5 \\ 0 & 1 & 10 & | & -28 \\ 0 & 0 & -7 & | & 21 \end{pmatrix}$

Therefore, $z = -3$
$y = -28 - 10z = -28 + 30 = 2$ and
$x = 5 - y/2 + z = 5 - 1 - 3 = 1.$

Much neater! And that is just one of an infinite number or reasons we use matrices!

Two For You – Solving Two Linear Equations Using Substitution

Solve the linear systems using the substitution method:

1) $x - 2y = 3$, $2x + y = 1$ 2) $4x + 3y = -1$, $7x + 2y = 8$

Answers 1) $x = 1$, $y = -1$ 2) $x = 2$, $y = -3$

Two For You – Solving Two Linear Equations

Solve the linear systems:

1) $5x - 7y = 10$, $15x + 4y = 5$ 2) $4x + 3y = -1$, $7x + 2y = 8$

Answers 1) $x = \dfrac{3}{5}$, $y = -1$ 2) $x = 2$, $y = -3$

Two For You – Solving Three Linear Equations

Solve the linear systems:
1) $x - 2y + z = 7$, $2x - y + 4z = 17$, $3x - 2y + 2z = 14$
2) $2x + y - 3z = 5$, $3x - 2y + 2z = 5$, $5x - 3y - z = 16$

Answers 1) $x = 2$, $y = -1$, $z = 3$ 2) $x = 1$, $y = -3$, $z = -2$

Two For You – Solving Three Linear Equations with MATRICES

Solve the linear systems using matrix notation:
1) $x - 2y + z = 7$, $2x - y + 4z = 17$, $3x - 2y + 2z = 14$
2) $2x + y - 3z = 5$, $3x - 2y + 2z = 5$, $5x - 3y - z = 16$

Answers 1) Start with the matrix $\begin{pmatrix} 1 & -2 & 1 & | & 7 \\ 2 & -1 & 4 & | & 17 \\ 3 & -2 & 2 & | & 14 \end{pmatrix}$. $x = 2$, $y = -1$, $z = 3$

2) Start with the matrix $\begin{pmatrix} 2 & 1 & -3 & | & 5 \\ 3 & -2 & 2 & | & 5 \\ 5 & -3 & -1 & | & 16 \end{pmatrix}$. $x = 1$, $y = -3$, $z = -2$

Consistent vs Inconsistent vs Dependent vs Unique Solutions of Three Linear Equations in 3 Unknowns

Example 1) For each of the following systems of equations, using **"row reduction"**, the system has been reduced so that the solution can be easily determined. State the solutions for each system.

(a) $\begin{array}{|l|} x + 2y - 6z = 5 \\ 4x + 5y - 21z = 5 \\ -3x + 3y + 17z = -2 \end{array}$ which can be reduced to $\begin{array}{|l|} x + 2y - 6z = 5 \\ y - z = 5 \\ z = -4 \end{array}$

(b) $\begin{array}{|l|} x - 2y + 4z = 2 \\ 2x - 3y + 5z = 3 \\ 3x - 4y + 6z = 7 \end{array}$ which can be reduced to $\begin{array}{|l|} x - 2y + 4z = 2 \\ y - 3z = -1 \\ 0 = 3 \end{array}$

(c) $\begin{array}{|l|} x + 2y + 3z = 3 \\ 2x + 3y + 8z = 4 \\ 3x + 2y + 17z = 1 \end{array}$ which can be reduced to $\begin{array}{|l|} x + 2y + 3z = 3 \\ y - 2z = 2 \\ 0 = 0 \end{array}$

Solution 1)(a) Unique solution. The three planes intersect in a single point: $z = -4$, $y = 1$, and $x = -21$.

(b) **Inconsistent solution.** The three planes have no common point of intersection. This can happen when at least two of the planes are parallel but not coincident.

(c) **Consistent (dependent) solution.** This happens when at least two of the planes coincide or no two of the three are parallel but they have a common line of intersection. If all three planes coincide, you obtain only one non-trivial equation, that is one non-"$0 = 0$" equation. Here, with "free" variable (or "parameter") z, we have $y = 2 + 2z$ and so $x = 3 - 3z - 2(2 + 2z) = 3 - 3z - 4 - 4z = -1 - 7z$.

Solving Quadratic Equations Using the Quadratic Formula

Some quadratics factor and solve very easily, such as $x^2 - 4x + 3 = (x-3)(x-1) = 0$. Others, such as $2x^2 - x - 5 = 0$, have "less pleasant" real roots. Still others, such as $2x^2 - x + 3 = 0$, have non-real roots. In these latter cases, the quadratic formula makes solving for x much simpler.

> **Remember the quadratic formula**: if $ax^2 + bx + c = 0$, then $x = \dfrac{-b \pm \sqrt{b^2 - 4ac}}{2a}$.
>
> **Note the relationship between finding the roots and factoring the expression:**
>
> $ax^2 + bx + c = a\left(x - \dfrac{-b + \sqrt{b^2 - 4ac}}{2a}\right)\left(x - \dfrac{-b - \sqrt{b^2 - 4ac}}{2a}\right)$

Example 1) Solve the equation $2x^2 - x - 5 = 0$ using the quadratic formula.

Solution $a = 2, \; b = -1, \; c = -5$

$\therefore \; x = \dfrac{-(-1) \pm \sqrt{(-1)^2 - 4(2)(-5)}}{2(2)} = \dfrac{1 \pm \sqrt{41}}{4}$

The roots are $x = \dfrac{1 + \sqrt{41}}{4}$ and $x = \dfrac{1 - \sqrt{41}}{4}$.

> **Note the relationship between finding the roots and factoring the expression:**
>
> $2x^2 - x - 5 = 2\left(x - \dfrac{1 + \sqrt{41}}{4}\right)\left(x - \dfrac{1 - \sqrt{41}}{4}\right)$

Example 2) Solve the equation $2x^2 - x + 3 = 0$ using the quadratic formula.

Solution $a = 2, \; b = -1, \; c = 3$

$\therefore \; x = \dfrac{-(-1) \pm \sqrt{(-1)^2 - 4(2)(3)}}{2(2)} = \dfrac{1 \pm \sqrt{-23}}{4} = \dfrac{1 \pm \sqrt{23}\, i}{4}$

The roots are $x = \dfrac{1 + i\sqrt{23}}{4}$ and $x = \dfrac{1 - i\sqrt{23}}{4}$.

> **Note the relationship between finding the roots and factoring the expression:**
>
> $2x^2 - x + 3 = 2\left(x - \dfrac{1 + i\sqrt{23}}{4}\right)\left(x - \dfrac{1 - i\sqrt{23}}{4}\right)$

Factoring Quadratic Expressions Using the Quadratic Formula

Some quadratic expressions factor very easily, such as $x^2 - 4x + 3 = (x-3)(x-1)$. When we need to factor quadratic expressions whose corresponding equations have "unpleasant" real roots, such as $2x^2 - x - 5 = 0$, or non-real roots, such as $2x^2 - x + 3 = 0$, we can use the quadratic formula to make factoring simple.

Remember the quadratic formula: if $ax^2 + bx + c = 0$, then $x = \dfrac{-b \pm \sqrt{b^2 - 4ac}}{2a}$.

Here is a quadratic fact of life. Let $f(x) = ax^2 + bx + c$ have roots r_1 and r_2. Then we can factor $f(x) = ax^2 + bx + c = a(x - r_1)(x - r_2)$. **Note the "$a$" in the factored form.**

Example 1) Factor $f(x) = 2x^2 - x - 5$ using the quadratic formula.

Solution $a = 2$, $b = -1$, $c = -5$ $\therefore x = \dfrac{-(-1) \pm \sqrt{(-1)^2 - 4(2)(-5)}}{2(2)} = \dfrac{1 \pm \sqrt{41}}{4}$

The roots are $r_1 = \dfrac{1 + \sqrt{41}}{4}$ and $r_2 = \dfrac{1 - \sqrt{41}}{4}$

$\therefore f(x) \overset{a(x-r_1)(x-r_2)}{=} 2\left(x - \dfrac{1 + \sqrt{41}}{4}\right)\left(x - \dfrac{1 - \sqrt{41}}{4}\right) \overset{\text{or if you prefer...}}{=} \left(2x - \dfrac{1 + \sqrt{41}}{2}\right)\left(x - \dfrac{1 - \sqrt{41}}{4}\right)$

Example 2) Factor the expression $g(x) = 2x^2 - x + 3$ using the quadratic formula.

Solution $a = 2$, $b = -1$, $c = 3$ $\therefore x = \dfrac{-(-1) \pm \sqrt{(-1)^2 - 4(2)(3)}}{2(2)} = \dfrac{1 \pm \sqrt{-23}}{4} = \dfrac{1 \pm \sqrt{23}\,i}{4}$

The roots are $r_1 = \dfrac{1 + i\sqrt{23}}{4}$ and $r_2 = \dfrac{1 - i\sqrt{23}}{4}$

$\therefore g(x) \overset{a(x-r_1)(x-r_2)}{=} 2\left(x - \dfrac{1 + i\sqrt{23}}{4}\right)\left(x - \dfrac{1 - i\sqrt{23}}{4}\right) \overset{\text{or if you prefer...}}{=} \left(2x - \dfrac{1 + i\sqrt{23}}{2}\right)\left(x - \dfrac{1 - i\sqrt{23}}{4}\right)$

Problems Involving the Sum and Product of the Roots of a Quadratic Equation

Remember the quadratic formula: if $ax^2 + bx + c = 0$, then $x = \dfrac{-b \pm \sqrt{b^2 - 4ac}}{2a}$.

Adding the two roots:

$$r_1 + r_2 = \frac{-b + \sqrt{b^2 - 4ac}}{2a} + \frac{-b - \sqrt{b^2 - 4ac}}{2a} = \frac{-2b}{2a} = -\frac{b}{a}$$

Multiplying the two roots:

$$r_1 r_2 = \left(\frac{-b + \sqrt{b^2 - 4ac}}{2a}\right)\left(\frac{-b - \sqrt{b^2 - 4ac}}{2a}\right) \underset{\text{to a difference of squares!}}{\overset{\text{The numerator expands}}{=}} \frac{b^2 - (b^2 - 4ac)}{4a^2} = \frac{4ac}{4a^2} = \frac{c}{a}$$

Example 1) Identify the sum and product of the roots of $3x^2 - 4x - 2 = 0$ **without solving for the roots!**

Solution $a = 3$, $b = -4$, $c = -2$

$$\therefore \quad r_1 + r_2 \overset{\boxed{\text{Sum} = -\frac{b}{a}}}{=} -\frac{(-4)}{3} = \frac{4}{3} \quad \text{and} \quad r_1 r_2 \overset{\boxed{\text{Product} = \frac{c}{a}}}{=} -\frac{2}{3}$$

Example 2) The sum and product of the roots of a quadratic equation are $\dfrac{4}{3}$ and $-\dfrac{2}{3}$, respectively. Find the quadratic equation.

Solution Rewrite $ax^2 + bx + c = 0$ as $x^2 + \dfrac{b}{a}x + \dfrac{c}{a} = 0 = x^2 - \left(-\dfrac{b}{a}\right)x + \dfrac{c}{a}$.

We are given $r_1 + r_2 = \dfrac{4}{3} = -\dfrac{b}{a}$ and $r_1 r_2 = -\dfrac{2}{3} = \dfrac{c}{a}$.

Therefore, the required quadratic equation is

$$x^2 - \left(\frac{4}{3}\right)x - \frac{2}{3} = 0 \quad \text{or} \quad 3x^2 - 4x - 2 = 0.$$

Four For You – Consistent vs Inconsistent vs Dependent vs Unique Solutions of Three Linear Equations in 3 Unknowns

The following equations are the reduced forms of four linear systems in three variables. State the solution for each and whether it is dependent, inconsistent, and/or unique.

1) $x+y+z=3$, $y-z=5$, $z=1$
2) $x+y+z=3$, $y-z=5$, $0=0$
3) $x+y+z=3$, $0=0$, $0=0$
4) $x+y+z=3$, $y-z=5$, $0=1$

Answers 1) $z=1$, $y=6$, $x=-4$; unique solution

2) $z \in \mathbb{R}$, $y=5+z$, $x=-2-2z$; dependent system with infinite solutions (one free variable)

3) $z \in \mathbb{R}$, $y \in \mathbb{R}$, $x=3-y-z$; dependent system with infinite solutions (two free variables)

4) no solution; inconsistent.

Two For You – Solving Quadratics Using the Quadratic Formula

Solve each of the following using the quadratic formula:

1) $3x^2 - 4x - 1 = 0$ 2) $5x^2 + 6x + 5 = 0$

Answers 1) $\dfrac{2+\sqrt{7}}{3}$, $\dfrac{2-\sqrt{7}}{3}$ 2) $\dfrac{-3+4i}{5}$, $\dfrac{-3-4i}{5}$

Two For You – Factoring Using the Quadratic Formula

Factor each of the following using the quadratic formula:

1) $3x^2 - 4x - 1$ 2) $5x^2 + 6x + 5$

Answers 1) $3\left(x - \dfrac{2+\sqrt{7}}{3}\right)\left(x - \dfrac{2-\sqrt{7}}{3}\right) \overset{\text{or if you prefer...}}{=} (3x - 2 - \sqrt{7})\left(x - \dfrac{2-\sqrt{7}}{3}\right)$

2) $5\left(x - \dfrac{-3+4i}{5}\right)\left(x - \dfrac{-3-4i}{5}\right) \overset{\text{or if you prefer...}}{=} (5x + 3 - 4i)\left(x + \dfrac{3+4i}{5}\right)$

Two For You – Sum and Product of the Roots of a Quadratic

1) Identify the sum and product of the roots of $\pi x^2 + ex - 1 = 0$ **without solving for the roots!**

2) The sum and product of the roots of a quadratic equation are $-\dfrac{2}{5}$ and 3, respectively. Find the quadratic equation.

Answers 1) Sum $= -\dfrac{e}{\pi}$, Product $= -\dfrac{1}{\pi}$

2) $x^2 + \dfrac{2}{5}x + 3 = 0$ or $5x^2 + 2x + 15 = 0$

The Graph of $y = a(x-b)^2 + c$

Given the parabola $y = a(x-b)^2 + c$, the vertex is (b, c) and the graph opens up if $a > 0$ and down if $a < 0$. The y intercept (where $x = 0$) is $ab^2 + c$.

Example 1) State the vertex and y intercept and draw the graph for each of the following: (a) $y = (x-2)^2 + 1$ (b) $y = -2(x+2)^2 + 4$

Solution

(a) vertex: $(2, 1)$; y intercept $= 5$ (b) vertex: $(-2, 4)$; y intercept $= -4$

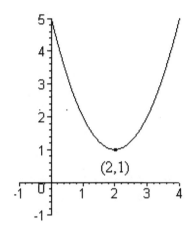
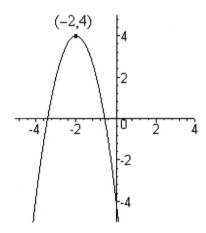

Example 2) Let $y = x - x^2$. Find the vertex and draw the graph.

Solution To find the vertex, **complete the square**.

$$y = x - x^2 = -(x^2 - x) \underset{\text{The coefficient of } x \text{ is } -1.\text{ Divide by 2 and square!}}{=} -\left(x^2 - x + \frac{1}{4}\right) + \frac{1}{4} \; \overset{\text{We subtracted 1/4 so we add 1/4.}}{} = -\left(x - \frac{1}{2}\right)^2 + \frac{1}{4}$$

The vertex is $\left(\dfrac{1}{2}, \dfrac{1}{4}\right)$

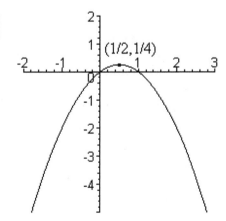

Completing the Square

Let's set this up so that we always have $a > 0$:

$(x+a)^2 = x^2 + 2ax + a^2 \qquad (x-a)^2 = x^2 - 2ax + a^2$

Note the sequence a, $2a$, and a^2. Completing the square is easy!

1) **Identify the $2a$ term**.

2) **Divide by 2 to get the a term**.

3) **Square to get the a^2 term**.

Example 1) Complete the square in the expression $4x^2 + 12x - 17$.

Solution $4x^2 + 12x - 17$

[Factor the 4 from the x^2 term and the x term **but not the constant!**]

$= 4(x^2 + 3x) - 17$

[Now, $2a=3$, so $a=3/2$ and $a^2=9/4$. By the way, note how we got "$3x$": $3x=12x/4$. Keep this in mind for the **next** example.]

$= 4(x^2 + 3x + \frac{9}{4}) - 17 - 4(9/4)$ $\boxed{\text{We add } 4\left(\frac{9}{4}\right) \text{ so we subtract } 4\left(\frac{9}{4}\right).}$

$= 4\left(x + \frac{3}{2}\right)^2 - 26$

Always factor out the coefficient of x^2. Factor it from the x term as well but **NOT** the constant.

Example 2) Complete the square in the expression $5 - \frac{3}{2}x^2 + 2x$.

Solution $5 - \frac{3}{2}x^2 + 2x$ $\quad$ [Factor out $-\frac{3}{2}$. Note: $-\frac{4}{3}x = \frac{2x}{\left(-\frac{3}{2}\right)}$] $= -\frac{3}{2}\left(x^2 - \frac{4}{3}x\right) + 5$

$= -\frac{3}{2}\left(x^2 - \frac{4}{3}x + \frac{4}{9}\right) + 5 + \frac{3}{2}\left(\frac{4}{9}\right)$ $\boxed{\begin{array}{l}\text{Note } 2a = -4/3, \text{ so } a = -2/3, \text{ and } a^2 = 4/9. \\ \text{Also, we \textbf{SUBTRACT} } \frac{3}{2}\left(\frac{4}{9}\right) \text{ so,} \\ \text{to compensate, we \textbf{ADD} } \frac{3}{2}\left(\frac{4}{9}\right).\end{array}}$

$= -\frac{3}{2}\left(x - \frac{2}{3}\right)^2 + 5 + \frac{2}{3} = -\frac{3}{2}\left(x - \frac{2}{3}\right)^2 + \frac{17}{3}$

Solving Linear Inequalities

1) One of math's cardinal rules: **what you do to one side you do to the other!**

2) Another: when you multiply or divide an **inequality** by a **negative**, the direction of the inequality **reverses**.

> For example, $-2 < 4$. Multiply both sides by -3 and you get $6 \underset{\text{The direction is reversed!}}{>} -12$.

Example 1) Solve the inequality $4 - 2x < 5 + 8x$.

Solution $4 - 2x < 5 + 8x$

Bring the x terms to the left and the numbers to the right: $-10x < 1$

Now divide both sides by -10: $x \underset{\text{We divided by a negative so the inequality reverses!}}{>} -\dfrac{1}{10}$

OR

$4 - 2x < 5 + 8x$

Bring the numbers to the left and the x terms to the right: $-1 < 10x$

Now divide both sides by 10: $-\dfrac{1}{10} \underset{\text{This time the inequality direction DID NOT CHANGE because we divided by a POSITIVE number!}}{<} x$

Example 2) Solve the inequality $3 \leq 2x - 5 < 7$.

Solution $3 \leq 2x - 5 < 7 \quad \underset{\text{Add 5 to each of the three parts of the inequality...}}{\Leftrightarrow} \quad 8 \leq 2x < 12 \quad \underset{\text{...and divide by 2.}}{\Leftrightarrow} \quad 4 \leq x < 6$

Example 3) Solve the inequality $3 - 2x < 6 + 4x < 7$.

Solution This time we **MUST** solve two inequalities separately, because x appears more than once.

$\begin{array}{lll}
3 - 2x < 6 + 4x & \text{and} & 6 + 4x < 7 \\
-6x < 3 & \text{and} & x < 1 \\
x > -\dfrac{1}{2} & \text{and} & x < \dfrac{1}{4} \quad \therefore \; -\dfrac{1}{2} < x < \dfrac{1}{4}
\end{array}$

We need to say "**and**" because **both** inequalities must be satisfied!

Solving Quadratic Inequalities

Some of these inequalities factor and solve very easily. Some don't factor, which means there are no intercepts: the parabola is either always above or always below the x axis. Some factor if you first **complete the square** and then use **difference of squares**.

Example 1) Solve the inequality $x^2 - 3x - 4 > 0$.

Solution $x^2 - 3x - 4 > 0 \Leftrightarrow (x-4)(x+1) > 0$
The solution is $x \in (-\infty, -1) \cup (4, \infty)$.

$x+1 < 0$	$x+1 > 0$	$x+1 > 0$
$x-4 < 0$	$x-4 < 0$	$x-4 > 0$
$(-)(-)$	$(+)(-)$	$(+)(+)$

$$\underset{+}{} \underset{-1}{\mid} \underset{-}{} \underset{4}{\mid} \underset{+}{}$$

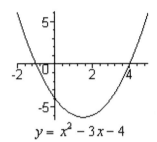
$y = x^2 - 3x - 4$

Example 2) Solve the inequality $x^2 - x + 1 < 0$.

Solution $x^2 - x + 1 < 0$

[Completing the square: $(x+a)^2 = x^2 + 2ax + a^2$. Here, $a = -\frac{1}{2}$.]

$\Leftrightarrow x^2 - x + \frac{1}{4} + 1 - \frac{1}{4} < 0 \Leftrightarrow \left(x - \frac{1}{2}\right)^2 + \frac{3}{4} < 0$

Since $\left(x - \frac{1}{2}\right)^2 + \frac{3}{4} \geq \frac{3}{4} \overset{\text{Always!}}{>} 0$, there is no solution.

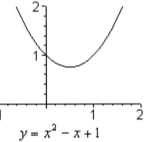
$y = x^2 - x + 1$

Example 3) Solve the inequality $x^2 - x - 1 < 0$.

Solution $x^2 - x - 1 < 0 \Leftrightarrow x^2 - x + \frac{1}{4} - 1 - \frac{1}{4} < 0 \Leftrightarrow \left(x - \frac{1}{2}\right)^2 - \frac{5}{4} < 0$

[$a^2 - b^2 = (a-b)(a+b)$, $a = x - \frac{1}{2}$, $b = \frac{\sqrt{5}}{2}$]

[Be careful with "−" when you make "2" a common denominator for each root.]

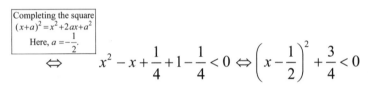

$$\underset{+}{} \underset{\frac{1-\sqrt{5}}{2}}{\mid} \underset{-}{} \underset{\frac{1+\sqrt{5}}{2}}{\mid} \underset{+}{}$$

$\therefore x \in \left(\frac{1-\sqrt{5}}{2}, \frac{1+\sqrt{5}}{2}\right)$

Two For You – The Graph of $y = a(x-b)^2 + c$

Find the vertex, the y intercept, and state whether the parabola opens up or down:

1) $y = 2(x+3)^2 - 5$ 2) $y = 1 - 8x - 2x^2$

Answers 1) vertex: $(-3, -5)$; y intercept $= 13$; up
2) vertex: $(-2, 9)$; y intercept $= 1$; down

Two For You – Completing the Square

Complete the square for each of the following:

1) $3x^2 - 30x - 11$ 2) $x - x^2$

Answers 1) $3(x-5)^2 - 86$ 2) $\dfrac{1}{4} - \left(x - \dfrac{1}{2}\right)^2$

Two For You – Solving Linear Inequalities

Solve these inequalities:

1) $3 > 5 - 6x \geq 2$ 2) $3x + 7 < 4x + 5 < -x + 5$

Answers 1) $\dfrac{1}{3} < x \leq \dfrac{1}{2}$

2) No solution.

$(3x + 7 < 4x + 5 \Leftrightarrow x > 2$ while $4x + 5 < -x + 5 \Leftrightarrow x < 0$.

There are no numbers than are greater than 2 **AND** less than 0!)

Two For You – Solving Quadratic Inequalities

Solve the inequalities: 1) $x^2 + 7x + 12 > 0$ 2) $x^2 + 2x - 1 < 0$

Answers 1) $x \in (-\infty, -4) \cup (-3, \infty)$ 2) $x \in \left(-1 - \sqrt{2}, -1 + \sqrt{2}\right)$

Solving Inequalities with Two or More Factors

Here, you will have an inequality where

1) on one side you will have a **product** and/or **quotient** of factors in the form $x \pm a$;

2) on the other side **you must have 0.**

Example 1) Solve the inequality $(x+3)x(x-4) < 0$.

Solution

```
      (-)(-)(-)     (+)(-)(-)     (+)(+)(-)     (+)(+)(+)
  ———————+—————————————+—————————————+—————————————
    -    -3      +      0      -      4      +
```

$\therefore x \in (-\infty, -3) \cup (0, 4)$

Example 2) Solve the inequality $(x+3)^2(x+1)^3 x^{1/3}(x-4) \leq 0$.

Solution This time, the factor $(x+3)$ **DOESN'T MATTER** because it is raised to an **EVEN** exponent: $(x+3)^2 \geq 0$ **ALWAYS**. I will include -3 on the number line just for emphasis! However, $x+1$ is raised to an **ODD** exponent and so it will change from $-$ to $+$ as x goes from less than -1 to greater than -1. Ditto for the $x^{1/3}$ term as x goes from less than 0 to greater than 0.

```
  (-)(-)(-)  (-)(-)(-)  (+)(-)(-)     (+)(+)(-)     (+)(+)(+)
  ———————+—————————+—————————+—————————————+—————————————
    -    -3    -    -1    +    0      -      4      +
```

The solution is $x \in (-\infty, -1] \cup [0, 4]$.

Example 3) Solve the inequality $\dfrac{(x+3)(x+1)x}{(x-4)} \leq 0$.

Solution Watch out for division by 0: we **can't** let x be 4.

```
  (-)(-)(-)(-)  (+)(-)(-)(-)  (+)(+)(-)(-)     (+)(+)(+)(-)     (+)(+)(+)(+)
  ——————————+———————————+———————————+———————————————+———————————————
    +      -3      -      -1      +      0      -      4      +
```

$\therefore x \in [-3, -1] \cup [0, 4)$

Solving Rational Inequalities

In these questions, if you **cross multiply**, you need **SEPARATE CASES**.

Multiply by a " + " and the direction stays the same!
Multiply by a " − " and the direction reverses!

HERE IS AN EASIER WAY...

Example 1) Solve the inequality $\dfrac{1}{x+2} \leq \dfrac{2}{3x+1}$.

Solution $\dfrac{1}{x+2} \leq \dfrac{2}{3x+1}$ [Do not cross-mulitply!] $\Leftrightarrow \dfrac{1}{x+2} - \dfrac{2}{3x+1} \leq 0$

[Get a common denominator.] $\Leftrightarrow \dfrac{3x+1-2(x+2)}{(x+2)(3x+1)} \leq 0$ [Simplify.] $\Leftrightarrow \dfrac{x-3}{3(x+2)\left(x+\dfrac{1}{3}\right)} \leq 0$

```
      (−)(−)(−)     (+)(−)(−)     (+)(+)(−)     (+)(+)(+)
  ─────────────┼──────────┼──────────┼──────────────
       −       −2    +    −1/3   −     3    +
```

The solution is $x \in (-\infty, -2) \cup \left(-\dfrac{1}{3}, 3\right]$.

Example 2) Solve the inequality $\dfrac{2x-1}{3x+1} \geq \dfrac{x+2}{x-2}$.

Solution $\dfrac{2x-1}{3x+1} \geq \dfrac{x+2}{x-2}$ [Don't cross multiply!] $\Leftrightarrow \dfrac{2x-1}{3x+1} - \dfrac{x+2}{x-2} \geq 0$

[common denominator] $\Leftrightarrow \dfrac{(2x-1)(x-2) - (x+2)(3x+1)}{(3x+1)(x-2)} \geq 0$ [Expand...] $\Leftrightarrow \dfrac{2x^2 - 5x + 2 - (3x^2 + 7x + 2)}{(3x+1)(x-2)} \geq 0$

[...and simplify.] $\Leftrightarrow \dfrac{-x^2 - 12x}{(3x+1)(x-2)} \geq 0$ [Multiply both sides by −1.] $\Leftrightarrow \dfrac{x^2 + 12x}{(3x+1)(x-2)} \leq 0$ [Factor just a little more.] $\Leftrightarrow \dfrac{x(x+12)}{3\left(x+\dfrac{1}{3}\right)(x-2)} \leq 0$

```
   (−)(−)(−)(−)   (+)(−)(−)(−)   (+)(+)(−)(−)   (+)(+)(+)(−)   (+)(+)(+)(+)
  ──────────┼──────────┼──────────┼──────────┼──────────
      +    −12    −   −1/3    +       0    −     2    +
```

Therefore, $x \in \left[-12, -\dfrac{1}{3}\right) \cup [0, 2)$.

The Basics of Absolute Value

Remember, absolute value is **ALWAYS NON-NEGATIVE**!

$|3| = 3$, $\quad |-3| = 3 \underset{\substack{\text{This step is}\\\text{THE KEY STEP}\\\text{for understanding}\\\text{absolute value!}}}{=} -(-3)$ and so $|x| = \begin{cases} -x, & \text{if } x < 0 \\ x, & \text{if } x \geq 0 \end{cases}$

Here is the part that people find so confusing.

WHY PUT "−" IN FRONT OF THE x?

Why not just make it positive as we did with -3?

BECAUSE ONE OF THE "−" SIGNS IS INSIDE THE x, so you can't get rid of it!

$f(x) = |x| = \begin{cases} -x, & \text{if } x < 0 \\ x, & \text{if } x \geq 0 \end{cases}$

Domain $= \mathbb{R}$, Range $= [0, \infty)$

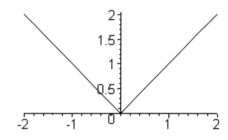

Example 1) Write (a) $|x - 1|$ (b) $|x + 1|$ without using absolute value notation.

Solution (a) $|x - 1| = \begin{cases} -(x - 1), & \text{if } x < 1 \\ x - 1, & \text{if } x \geq 1 \end{cases}$ (b) $|x + 1| = \begin{cases} -(x + 1), & \text{if } x < -1 \\ x + 1, & \text{if } x \geq -1 \end{cases}$

Example 2) Graph: (a) $y = |x - 1|$ (b) $y = |x + 1|$.

Solution

(a) $y = |x - 1|$

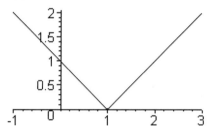

(b) $y = |x + 1|$

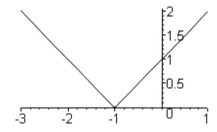

Solving Absolute Value Equations

Remember, absolute value is **ALWAYS NON-NEGATIVE**

Example 1) Solve the following absolute value equations.
(a) $|x|=4$ (b) $|x|=-3$ (c) $|2x-5|=7$
(d) $|x+5|=2x-3$ (e) $|2x+1|=|x-7|$

Solution (a) $x=4$ or $x=-4$

(b) No solution since $|x|$ is positive.

(c) Either $2x-5=7$ in which case $2x=12$ and $x=6$
or
$2x-5=-7$ in which case $2x=-2$ and $x=-1$.
(Remember, x can be negative; it is the absolute value of x that must be positive.)

(d) Easiest method here is to use cases.

Case 1) $x+5 \geq 0$ so $x \geq -5$. In this case, $|x+5|=x+5$. So we solve:
$x+5=2x-3 \Leftrightarrow -x=-8 \Leftrightarrow x=8$. Since $8 \geq -5$, $x=8$ is a solution.

Case 2) $x+5<0$ so $x<-5$. In this case, $|x+5|=-(x+5)$. So we solve:
$-(x+5)=2x-3 \Leftrightarrow -x-5=2x-3 \Leftrightarrow -3x=2 \Leftrightarrow x=-\frac{2}{3}$
Since $-\frac{2}{3}$ **IS NOT LESS THAN** -5, $x=-\frac{2}{3}$ is **NOT** a solution.

(e) Since $|a|=|b| \Leftrightarrow a^2=b^2$, the easiest method here is to **square both sides**.
$|2x+1|=|x-7| \Leftrightarrow (2x+1)^2=(x-7)^2 \Leftrightarrow 4x^2+4x+1=x^2-14x+49$
$\Leftrightarrow 3x^2+18x-48=0 \Leftrightarrow x^2+6x-16=0 \Leftrightarrow (x-2)(x+8)=0 \Leftrightarrow x=2$ or $x=-8$

We **DON'T** have to check our answers in this example because of "$\Leftrightarrow$"! Not only does each step follow from the previous step, each step is **REVERSIBLE**!
In example (d), if you use the method of squaring both sides, **YOU DO HAVE TO CHECK BECAUSE THE SQUARING IN THAT EXAMPLE IS NOT REVERSIBLE!**

Two For You – Solving Inequalities with Two or More Factors

Solve these inequalities:

1) $(2x-3)(4-x)(x-7)^3(x+1)^2 > 0$

$\left(\text{Hint: write the inequality as } -2\left(x-\dfrac{3}{2}\right)(x-4)(x-7)^3(x+1)^2 > 0 \text{ and then} \right.$
$\left. \left(x-\dfrac{3}{2}\right)(x-4)(x-7)^3(x+1)^2 < 0. \right)$

2) $\dfrac{(x+3)^2(x-1)}{(x+5)^{3/2}(x-2)(x-4)^3} \leq 0$ (Hint: $x > -5$; otherwise, $1/(x+5)^{3/2}$ is undefined.)

Answers 1) $x \in \left(-\infty, \dfrac{3}{2}\right) \cup (4,7)$ 2) $x \in (-5, 1] \cup (2, 4)$

Two For You – Solving Rational Inequalities

Solve: 1) $\dfrac{5}{x+7} \geq \dfrac{2}{x-5}$ 2) $\dfrac{x-1}{x+1} \leq \dfrac{3x-1}{3x+1}$

Answers 1) $(-7, 5) \cup [13, \infty)$ 2) $\left(-1, -\dfrac{1}{3}\right) \cup [0, \infty)$

Two For You – The Basics of Absolute Value

1) Write $|2x-3|$ without absolute value signs.
2) Graph $y=|2x-3|$.

Answers 1) $|2x-3| = \begin{cases} -(2x-3), & \text{if } x < \frac{3}{2} \\ (2x-3), & \text{if } x \geq \frac{3}{2} \end{cases}$ 2)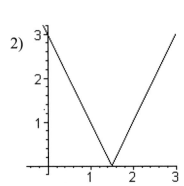

Three For You – Solving Absolute Value Equations

Solve: 1) $|x^3| = 1000$ 2) $2x+1 = |3x-2|$ 3) $|3x+2| = |x-6|$

Answers 1) $x=10$ or $x=-10$ 2) $x=3$ 3) $x=-4$ or $x=1$

Solving Easy Absolute Value Inequalities

Keep in mind that for $a > 0$:

$\boxed{|x| < a \Leftrightarrow -a < x < a}$ and $\boxed{|x| > a \Leftrightarrow x < -a \text{ OR } x > a}$

Example 1) Solve the following absolute value inequalities.

(a) $|x| < 1$ (b) $|x| \geq 4$ (c) $|x| < -1$ (d) $|x| > -2$

Solution (a) $|x| < 1 \Leftrightarrow -1 < x < 1$

(b) $|x| \geq 4 \Leftrightarrow x \leq -4 \text{ or } x \geq 4$

(c) $|x| < -1$ is **ALWAYS FALSE**, since $|x| \geq 0$.

(d) $|x| > -2 \Leftrightarrow x \in (-\infty, \infty)$, that is, $|x| > -2$ is **ALWAYS TRUE!**

Example 2) Solve: a) $|2x - 4| < 6$ (b) $|x - 3| \geq 7$

Solution (a) $|2x - 4| < 6 \Leftrightarrow -6 < 2x - 4 < 6$
$\Leftrightarrow -2 < 2x < 10$
$\Leftrightarrow -1 < x < 5, \text{ that is, } x \in (-1, 5)$

(b) $|x - 3| \geq 7 \Leftrightarrow x - 3 \leq -7 \text{ or } x - 3 \geq 7$
$\Leftrightarrow x \leq -4 \text{ or } x \geq 10, \text{ that is, } x \in (-\infty, -4] \cup [10, \infty)$

Solving Less Easy Absolute Value Inequalities

There are two basic methods to use when solving more complicated absolute value inequalities (and equations as well): "**cases**" or "**squaring both sides**". The key difference:

$|a| < b \quad \underset{\text{The implication goes in only one direction!}}{\Rightarrow} \quad a^2 < b^2.$

Note that here, since $|a| \geq 0$, b **MUST BE POSITIVE!**

When you square, you **could** introduce solutions for $a^2 < b^2$ that don't work for $|a| < b$. That's why you use cases in this type of problem.

However, $|a| < |b| \quad \underset{\text{The implication goes in both directions!}}{\Leftrightarrow} \quad a^2 < b^2$. Squaring both sides is faster than cases and solutions work both ways! BUT...

BE CAREFUL $a < |b| \not\Rightarrow a^2 < b^2$: a^2 can be less than b^2, **but doesn't have to be!**

Example 1) Solve $|x+3| < 2x$.

Solution Use cases since squaring both sides can lead to "extraneous" solutions.

Case 1) $x + 3 \geq 0$ so that $x \geq -3$. In this case, $|x+3| = x+3$
The inequality becomes $x + 3 < 2x$ and so $3 < x$. The solution in this case is $x > 3$.

Case 2) $x + 3 < 0$ so that $x < -3$. In this case, $|x+3| = -(x+3)$
The inequality becomes $-x - 3 < 2x$. Therefore, $-3 < 3x$ and so $-1 < x$.
Since $x > -1$ and $x < -3$ are incompatible, there are no solutions in this case.
Combining the two cases, the solution of $|x+3| < 2x$ is $x \in (3, \infty)$.

Example 2) Solve $|x+4| < |2x-6|$.

Solution In this example, squaring both sides is the best method.
(Note that there would be **four** separate cases if we used the case method.)

$|x+4| < |2x-6| \Leftrightarrow x^2 + 8x + 16 < 4x^2 - 24x + 36 \Leftrightarrow 0 < 3x^2 - 32x + 20$

(Personal preference: I like having the expression on the left and 0 on the right.)

$\Leftrightarrow 3x^2 - 32x + 20 > 0 \Leftrightarrow (3x-2)(x-10) > 0 \Leftrightarrow \left(x - \frac{2}{3}\right)(x-10) > 0$

```
      (-)(-)        (+)(-)         (+)(+)     ∴ x ∈ (-∞, 2/3) ∪ (10, ∞)
   ───────────┬──────────────┬──────────────
        +     2/3     -      10      +
```

The Basics of $\sqrt{\ }$ and the Reason $\sqrt{x^2} = |x|$

Example 1) State the domain and range of the function $y = \sqrt{x}$ and draw the graph.

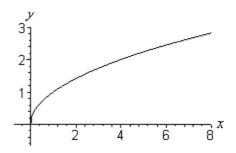

Solution The domain and range are both $[0, \infty)$.

Example 2) Solve: (a) $x = \sqrt{9}$ (b) $x = -\sqrt{9}$ (c) $x^2 = 9$

Solution (a) $x = \sqrt{9} = 3$ (b) $x = -\sqrt{9} = -3$ (c) $x^2 = 9 \Leftrightarrow x = 3$ or $x = -3$

Example 3) Evaluate $\sqrt{(3)^2}$, $\sqrt{(-3)^2}$, $|3|$, and $|-3|$. What does this tell you about $\sqrt{x^2}$?

Solution $\sqrt{(3)^2} = \sqrt{9} = 3$, $\sqrt{(-3)^2} = \sqrt{9} = 3$, $|3| = 3$, and $|-3| = 3$.

$\therefore \sqrt{x^2} = |x|$, for both $x > 0$ and $x < 0$. Put another way:

$$\sqrt{x^2} = |x| = \begin{cases} -x, & \text{if } x < 0 \\ x, & \text{if } x \geq 0 \end{cases}$$

For example:

when $x = 3$, we have $\sqrt{(3)^2} \overset{\substack{x>0 \text{ and so} \\ \sqrt{x^2}=x}}{=} 3 = |3|$

when $x = -3$, we have $\sqrt{(-3)^2} = \sqrt{9} = 3 \overset{\substack{x<0 \text{ and so} \\ \sqrt{x^2}=-x}}{=} -(-3) = |-3|$

Solving Equations Involving Square Roots

Here, as with the more complicated absolute value questions, squaring both sides works. BUT while $x = y \Rightarrow x^2 = y^2$, the converse is false: $x^2 = y^2 \not\Rightarrow x = y$. For example, $(4)^2 = (-4)^2$ but $4 \neq -4$! When we square both sides, we can introduce "**extraneous**" roots. We **must check** our possible solutions in the **original** equations!

Example 1) Solve $\sqrt{x+5} = x - 7$.

Solution $\sqrt{x+5} = x - 7 \Rightarrow x + 5 = x^2 - 14x + 49 \Rightarrow x^2 - 15x + 44 = 0$
$\Rightarrow (x-11)(x-4) = 0 \Rightarrow x = 11$ or $x = 4$.

Check $x = 11$ in the original equation:
Left Side $= \sqrt{11+5} = \sqrt{16} = 4$ Right Side $= 11 - 7 = 4$

Check $x = 4$ in the original equation:
Left Side $= \sqrt{4+5} = \sqrt{9} = 3$ Right Side $= 4 - 7 = -3$
Therefore, the only solution is $x = 11$.

Example 2) Solve $\sqrt{2x-7} - \sqrt{x-4} = 1$

Solution $\sqrt{2x-7} - \sqrt{x-4} = 1$

[Isolate one of the square roots.]
$\Rightarrow \quad \sqrt{2x-7} = \sqrt{x-4} + 1 \quad \Rightarrow \quad 2x - 7 = x - 4 + 2\sqrt{x-4} + 1$

[Isolate the remaining square root.]
$\Rightarrow \quad x - 4 = 2\sqrt{x-4}$

[Square both sides.]
$\Rightarrow \quad x^2 - 8x + 16 = 4x - 16 \Rightarrow x^2 - 12x + 32 = 0$
$\Rightarrow (x-4)(x-8) = 0 \Rightarrow x = 4$ or $x = 8$

Check $x = 4$ in the original equation:
Left Side $= \sqrt{2(4)-7} - \sqrt{4-4} = 1$ Right Side $= 1$

Check $x = 8$ in the original equation:
Left Side $= \sqrt{2(8)-7} - \sqrt{8-4} = 1$ Right Side $= 1$
Therefore, both 4 and 8 are solutions.

Two For You – Solving Easy Absolute Value Inequalities

1) Solve: (a) $|x| \leq 0.1$ (b) $|x| > 3.2$
2) Solve: (a) $|x+3| \leq 2$ (b) $1 < |3x-5|$

Answers 1)(a) $[-0.1, 0.1]$ (b) $(-\infty, -3.2) \cup (3.2, \infty)$
2)(a) $[-5, -1]$ (b) $\left(-\infty, \dfrac{4}{3}\right) \cup (2, \infty)$

Three For You – Solving Less Easy Absolute Value Inequalities

Solve: 1) $5x - 8 \geq |4 - x|$ (Hint: use $|4-x| = |x-4|$.)   2) $|3x-2| < |2x-3|$
3) Given an example showing that $a^2 < b^2 \not\Rightarrow |a| < b$.

Answers 1) $[2, \infty)$ 2) $(-1, 1)$
3) Let $a = 3$ and $b = -5$. Then $9 = a^2 < b^2 = 25$ but $3 = a \not< b = -5$.

Two For You – The Basics of $\sqrt{}$ and the Reason $\sqrt{x^2}=|x|$

1) State the domain and range of $f(x)=\sqrt{x+1}+1$ and draw the graph.

2) Write $y=\sqrt{(x-4)^2}$ using first absolute value and then a "branch" definition.

Answers

1) Domain $=[-1,\infty)$, Range $=[1,\infty)$

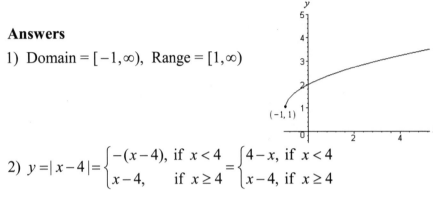

2) $y=|x-4|=\begin{cases}-(x-4),&\text{if }x<4\\x-4,&\text{if }x\geq 4\end{cases}=\begin{cases}4-x,&\text{if }x<4\\x-4,&\text{if }x\geq 4\end{cases}$

Two For You – Solving Equations Involving Square Roots

1) Solve: $\sqrt{2x-7}=x-3$

2) Solve: $\sqrt{x-3}-\sqrt{2x+1}=-2$ (Hint: first isolate the $\sqrt{2x+1}$ term.)

Answers 1) 4   2) 4, 12

Rationalizing Denominators that Have $\sqrt{}$

Often you run into problems where there is either a single term with a square root in the denominator or a binomial with one or two square roots. In the first case, a simple $\dfrac{\sqrt{}}{\sqrt{}}$ solves the problem. In the second, **DIFFERENCE OF SQUARES** comes to the rescue.

Example 1) Rationalize the denominators: (a) $\dfrac{3}{\sqrt{2}}$ (b) $\dfrac{\sqrt{7}}{2\sqrt{21}}$ (c) $\dfrac{xy}{\sqrt{2x}}$

Solution (a) $\dfrac{3}{\sqrt{2}} = \dfrac{3}{\sqrt{2}} \cdot \dfrac{\sqrt{2}}{\sqrt{2}} = \dfrac{3\sqrt{2}}{2}$

(b) $\dfrac{\sqrt{7}}{2\sqrt{21}} \overset{\text{Only the }\sqrt{21}\text{ is important!}}{=} \dfrac{\sqrt{7}}{2\sqrt{21}} \cdot \dfrac{\sqrt{21}}{\sqrt{21}} \overset{\sqrt{7}\sqrt{21}=\sqrt{7}\sqrt{7}\sqrt{3}}{=} \dfrac{\cancel{7}^{1}\sqrt{3}}{2(2\cancel{1}_{3})} = \dfrac{\sqrt{3}}{6}$

(c) $\dfrac{xy}{\sqrt{2x}} = \dfrac{xy}{\sqrt{2x}} \cdot \dfrac{\sqrt{2x}}{\sqrt{2x}} = \dfrac{\sqrt{2x}\,xy}{2x} = \dfrac{y\sqrt{2x}}{2}$

Example 2) Rationalize the denominators:

(a) $\dfrac{1}{\sqrt{x}-3}$ (b) $\dfrac{x}{\sqrt{x}+y}$ (c) $\dfrac{x}{\sqrt{2x+1}-3\sqrt{x-3}}$

Solution (a) $\dfrac{1}{\sqrt{x}-3} \overset{\substack{(a-b)(a+b)=a^2-b^2 \\ \text{Here, } a=\sqrt{x} \text{ and } b=3.}}{=} \left(\dfrac{1}{\sqrt{x}-3}\right)\left(\dfrac{\sqrt{x}+3}{\sqrt{x}+3}\right) = \dfrac{\sqrt{x}+3}{x-9}$

(b) $\dfrac{x}{\sqrt{x}+y} = \left(\dfrac{x}{\sqrt{x}+y}\right)\left(\dfrac{\sqrt{x}-y}{\sqrt{x}-y}\right) = \dfrac{x\sqrt{x}-xy}{x-y^2}$

(c) $\dfrac{x}{\sqrt{2x+1}-3\sqrt{x-3}} \overset{\substack{a=\sqrt{2x+1} \text{ and} \\ b=3\sqrt{x-3}}}{=} \left(\dfrac{x}{\sqrt{2x+1}-3\sqrt{x-3}}\right)\left(\dfrac{\sqrt{2x+1}+3\sqrt{x-3}}{\sqrt{2x+1}+3\sqrt{x-3}}\right)$

$= \dfrac{x\left(\sqrt{2x+1}+3\sqrt{x-3}\right)}{2x+1-9(x-3)} = \dfrac{x\left(\sqrt{2x+1}+3\sqrt{x-3}\right)}{-7x+28}$

Graphs of Basic Quadratic Relations

Four fundamental quadratic relations in variables x and y are the parabola, circle, ellipse, and hyperbola.

Example 1) Draw the graphs of the following relations.

(a) Parabola: $y = x^2$ (b) Circle: $x^2 + y^2 = 1$

(c) Ellipse: $\dfrac{x^2}{9} + \dfrac{y^2}{25} = 1$ (d) Hyperbola: $x^2 - y^2 = 1$

Solution

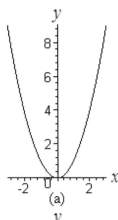
(a)

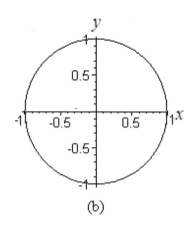
(b)

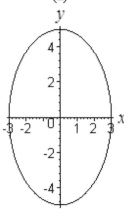
(c)

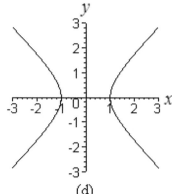
(d)

Notes:

(a) This parabola has vertex $(0,0)$ and opens upward.

(b) This circle has centre $(0,0)$ and radius 1.

(c) This ellipse has centre $(0,0)$ with x intercepts ± 3 and y intercepts ± 5.

(d) This hyperbola opens on the x axis with x intercepts ± 1. For comparison, note that the hyperbola $y^2 - x^2 = 1$ opens on the y axis with y intercepts ± 1.

Basic $y = x^n$ Graphs, where $n \in \mathbb{N}$ and Why Even and Odd Functions Are Called Even and Odd Functions

Graphs with equations of the form $y = x^n$, where $n \in \mathbb{N}$, come up so often that they deserve a special page. This is it!

Example 1) Draw the graphs of $y = x^3$ and $y = x^5$ on the same axes.

Solution

For **odd** natural numbers n:

1) $\lim\limits_{x \to \pm\infty} x^n = \pm\infty$ and the expression approaches $\pm\infty$ faster as n increases.

2) The graphs are symmetric in the **origin**, that is, $(-x)^n = -x^n$. This is the reason functions that satisfy $f(-x) = -f(x)$ are called **ODD** functions!

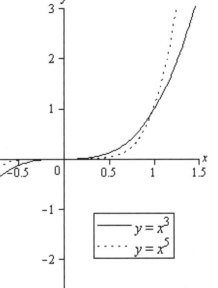

Example 2) Draw the graphs of $y = x^2$ and $y = x^4$ on the same axes.

Solution

For **even** natural numbers n:

1) $\lim\limits_{x \to \pm\infty} x^n = \infty$ and the expression approaches ∞ faster as n increases.

2) The graphs are symmetric in the y axis, that is, $(-x)^n = x^n$. This is the reason functions that satisfy $f(-x) = f(x)$ are called **EVEN** functions!

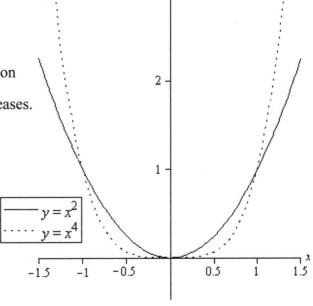

Basic $y = \dfrac{1}{x^n} = x^{-n}$ Graphs, where $n \in \mathbb{N}$

Graphs with equations of the form $y = \dfrac{1}{x^n}$, where $n \in \mathbb{N}$, come up so often that they deserve a special page. This is it!

Example 1) Draw the graphs of $y = \dfrac{1}{x}$ and $y = \dfrac{1}{x^3}$ on the same axes.

Solution

For **odd** natural numbers n:

1) $\lim\limits_{x \to \pm\infty} \dfrac{1}{x^n} = 0$ and the expression approaches 0 faster as n increases.

2) $\lim\limits_{x \to 0} \dfrac{1}{x^n} = \begin{cases} -\infty, & \text{if } x \to 0^- \\ \infty, & \text{if } x \to 0^+ \end{cases}$

and approaches $\pm\infty$ faster as n increases.

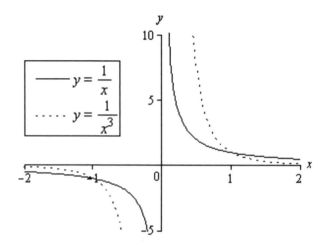

Example 2) Draw the graphs of $y = \dfrac{1}{x^2}$ and $y = \dfrac{1}{x^4}$ on the same axes.

Solution

For **even** natural numbers n:

1) $\lim\limits_{x \to \pm\infty} \dfrac{1}{x^n} = 0$ and the expression approaches 0 faster as n increases.

2) $\lim\limits_{x \to 0} \dfrac{1}{x^n} = \infty$ and approaches ∞ faster as n increases.

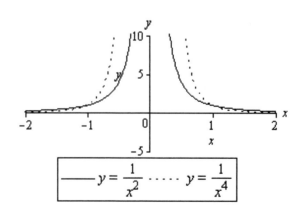

Two For You – Rationalizing Denominators that Have $\sqrt{}$

Rationalize the denominators: 1)(a) $\dfrac{\sqrt{3}}{\sqrt{45}}$ (b) $\dfrac{x-1}{\sqrt{x+4}}$ 2) $\dfrac{1}{\sqrt{x+h}-\sqrt{x}}$

Answers 1)(a) $\dfrac{\sqrt{15}}{15}$ (b) $\dfrac{(x-1)\sqrt{x+4}}{x+4}$ 2) $\dfrac{\sqrt{x+h}+\sqrt{x}}{h}$

One For You – Graphs of Basic Quadratic Relations

1) Graph each of the following quadratic relations and identify it as a parabola, circle, ellipse, or hyperbola.

(a) $y^2 - x^2 = 4$ (b) $y = 3 - 2x^2$ (c) $x^2 + y^2 = 9$ (d) $\dfrac{x^2}{4} + y^2 = 1$

Answers

1)(a) hyperbola

(b) parabola

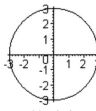
(c) circle

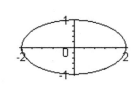
(d) ellipse

Two For You – Basic $y = x^n$ **Graphs, where** $n \in \mathbb{N}$

1) As $x \to \pm\infty$, which of these functions approach ∞ faster: $y = x^4$ or $y = x^6$?

2) As $x \to \pm\infty$, which of these functions approach $\pm\infty$ faster: $y = x^{11}$ or $y = x^9$?

Answers 1) $y = x^6$ 2) $y = x^{11}$

Two For You – Basic $y = \dfrac{1}{x^n} = x^{-n}$ **Graphs, where** $n \in \mathbb{N}$

1) As $x \to \pm\infty$, which of these functions approach 0 faster: $y = x^{-4}$ or $y = x^{-5}$?

2) As $x \to 0^-$, which of these functions approach $-\infty$ faster: $y = x^{-11}$ or $y = x^{-9}$?

Answers 1) $y = x^{-5}$ 2) $y = x^{-11}$

Basic $y = x^{\frac{1}{n}}$ Graphs, where $n \in \mathbb{N}$

Graphs with equations of the form $y = x^{\frac{1}{n}}$, where $n \in \mathbb{N}$, come up so often that they deserve a special page. This is it!

Example 1) Draw the graphs of $y = x^{\frac{1}{3}}$ and $y = x^{\frac{1}{5}}$ on the same axes.

Solution

For **odd** natural numbers n:

1) $\lim\limits_{x \to \pm\infty} x^{\frac{1}{n}} = \pm\infty$ and the expression approaches $\pm\infty$ **more slowly** as n increases.

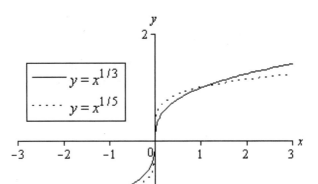

2) $\lim\limits_{x \to 0} x^{\frac{1}{n}} = 0$ and approaches 0 **more slowly** as n increases.

$$\left(\text{eg.,} \left(\frac{1}{64}\right)^{\frac{1}{3}} = \frac{1}{4} < \left(\frac{1}{64}\right)^{\frac{1}{6}} = \frac{1}{2} \right)$$

Example 2) Draw the graphs of $y = x^{\frac{1}{2}}$ and $y = x^{\frac{1}{4}}$ on the same axes.

Solution For **even** natural numbers n:

1) The domain is $[0, \infty)$.

2) $\lim\limits_{x \to \infty} x^{\frac{1}{n}} = \infty$ and the expression approaches ∞ **more slowly** as n increases.

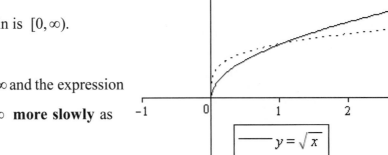

3) $\lim\limits_{x \to 0^+} x^{\frac{1}{n}} = 0$ and approaches 0 **more slowly** as n increases.

Shifting or Rescaling a Given Graph ($a > 0$)
$f(x+a), f(x-a), f(ax), af(x), f(x)+a, f(x)-a$

No matter what function $y = f(x)$ we begin with, the effect of each operation in the title of this section is **ALWAYS** the same.

Example 1) Let $y = f(x) = x^2$. Graph and describe these functions relative to $f(x)$.

(a) $y = f(x+2) = (x+2)^2$ (b) $y = f(x-2) = (x-2)^2$ (c) $y = f(2x) = (2x)^2 = 4x^2$

(d) $y = 2f(x) = 2x^2$ (e) $y = f(x) + 2 = x^2 + 2$ (f) $y = f(x) - 2 = x^2 - 2$

Solution

(a)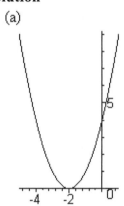
Shifts $f(x)$ 2 units left.

(b)
Shifts $f(x)$ 2 units right.

(c)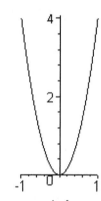
$f(x)$ values occur **twice** as fast.

(d)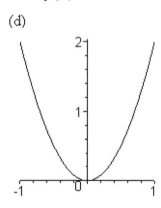
Doubles the $f(x)$ value.

(e)
Shifts $f(x)$ 2 units up.

(f)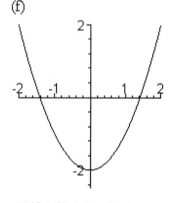
Shifts $f(x)$ 2 units down.

Note: I had to choose from one of two ways to present these graphs. First: keep the scaling the same on all. This would clearly illustrate the changing shapes in (c) and (d). However, it would make it more difficult to see how the translations in (a), (b), (e), and (f) change the original graph, since I would have had to take x from -4 to 4. Second: adjust the scaling while keeping the graphs the same size. This works well for translations but not so well for shape. I decided rescaling was the better solution overall.

Tests for Symmetry

Symmetry in the y axis: both (x, y) and $(-x, y)$ are on the graph.
Replace x with $-x$ and see if you obtain the same y value.
Symmetry in the x axis: both (x, y) and $(x, -y)$ are on the graph.
Replace y with $-y$ and see if you obtain the same x value.
Symmetry in the origin: both (x, y) and $(-x, -y)$ are on the graph.
Replace both x with $-x$ and y with $-y$ and see if you obtain the same x and y values.

Example 1) Test for symmetry in the following relations.
(a) $y = x^2$ (b) $y = x^3 + x$ (c) $x = \cos(y)$ (d) $x^2 + y^2 = 25$

Solution (a) y axis (replace x with $-x$): $(-x)^2 = x^2$; **symmetry in the y axis? YES**
x axis (replace y with $-y$): $-y = -x^2 \neq x^2$ for $x \neq 0$; **symmetry in the x axis? NO**
There is **no symmetry in the origin!** WHY? BECAUSE...

…you can have NO symmetry, ONE kind of symmetry, or ALL THREE.
But you <u>CANNOT</u> have exactly TWO KINDS OF SYMMETRY!*

(b) y axis: $(-x)^3 + (-x) = -x^3 - x \neq x^3 + x$, for $x \neq 0$. **NO!**
x axis: $-y = -(x^3 + x) \neq x^3 + x$. **NO!**
Origin: $(-x)^3 + (-x) = -x^3 - x = -(x^3 + x) = -y$, and so $x^3 + x = y$. **YES!**
(c) y axis: $-x = -\cos y$. **NO!**
x axis: $\cos(-y) = \cos y = x$. **YES!**
Origin: **NO!**
(d) y axis: $(-x)^2 + y^2 = x^2 + y^2 = 25$. **YES!**
x axis: $x^2 + (-y)^2 = x^2 + y^2 = 25$. **YES!**
Origin: **YES!**

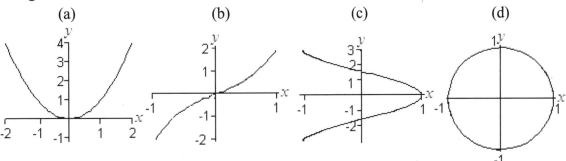

*When you have two kinds of symmetry, you **must** have three! Prove this!

Graphing Polynomials Without Calculus

Calculus tells us exactly where to find maximum and minimum points, the subtle changes at inflection points, and more. But with a little experience, we can tell a lot about the graph of a polynomial just from its equation.

Example 1) Determine the shape of the polynomial $P(x) = 3x^5 - x^4 + 2x - 5$, as x approaches ∞ and $-\infty$.

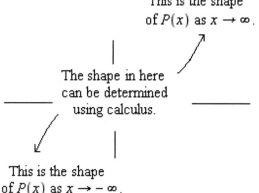

Solution $P(x)$ does what its "leading term", "$3x^5$", tells it to do when x is **BIG** in magnitude. So, no matter what is happening "in the middle"..............

If the polynomial is factored, we know its roots

$$\boxed{x - r \text{ is a factor} \Leftrightarrow x = r \text{ is a root}}$$

and how it "behaves" near the roots. If the exponent on the factor is **even**, the graph **doesn't change sign** as it passes through 0, that is, as it crosses the x axis at the root r. If the exponent is **odd, it does change sign**.

Example 2) Sketch the graph of $P(x) = x^2(x-1)^3$.

Solution The only roots are 0 and 1. From the exponents on the factors, we know $P(x) \leq 0$ for $x \leq 1$ and $P(x) \geq 0$ for $x \geq 1$. (The sign doesn't change at $x = 0$!) The leading term of $P(x)$, when expanded, is x^5. So, except for the "subtleties" of the curve, $P(x)$ looks like...

...this:

> The graph, like $y = x^5$, comes up from $-\infty$ as x does. It hits 0 at $x = 0$, heads down to a minimum (which we can find using CALCULUS!), comes back up to 0 at $x = 1$, and finally follows x^5 to ∞! The graph is curved but I used straight lines to show the tendencies. We find the **exact** shape of the curve (extremes, concavity) using derivatives.

Two For You – Basic $y = x^{\frac{1}{n}}$ Graphs, where $n \in \mathbb{N}$

1) As $x \to \infty$, which of these functions approach ∞ fastest:
$y = x^{1/4}$, $y = x^{1/5}$, or $y = x^{1/6}$?

2) As $x \to 0^-$, which of these functions approach 0 faster:
$y = x^{1/11}$ or $y = x^{1/9}$?

Answers 1) $y = x^{1/4}$ 2) $y = x^{1/9}$

One For You – Shifting or Rescaling a Given Graph

1) To the right is the graph of $y = f(x)$.
Sketch: (a) $y = f(x + 0.5)$ (b) $y = f(x - 0.5)$
(c) $y = f(0.5x)$ (d) $y = 0.5f(x)$

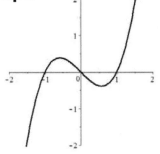

Answers 1) (a)   (b)   (c)   (d)

  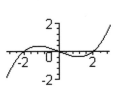 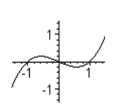

Two For You – Tests for Symmetry

Discuss the symmetry for the following relations: 1) $y = x\, e^{|x|}$ 2) $x^3 + y^3 = x$

Answers 1) origin   2) origin

Two For You – Graphing Polynomials without Calculus

Sketch the graphs without calculus:

1) $P(x) = (x-1)x^3(x+1)^2$ 2) $P(x) = -(x-1)^2 x^2 (x+1)^2$

Answers

1)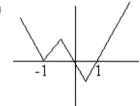

2)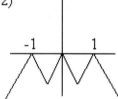

Note that
$P(x) = -(x-1)^2 x^2 (x+1)^2$
is symmetric in the y axis!

Vertical and Horizontal Asymptotes

A **vertical asymptote** is a **finite** x value $\boxed{x = \text{constant}}$ where $y \to \infty$ or $-\infty$. So, look for x values that make y "blow up": division by 0 or functions that have "built in" vertical asymptotes such as ln, tan, cot, sec, and csc.

A **horizontal asymptote** is a **finite** y value $\boxed{y = \text{constant}}$ where $x \to \infty$ or $-\infty$. Take the limit as $x \to \pm\infty$ to see if you obtain a finite y.

Example 1) Find the vertical and horizontal asymptotes for the function
$$y = \frac{1}{(x-5)(x+1)} + \ln(x-1)$$

Solution $x = 5$ is a vertical asymptote. Also,
$$\lim_{x \to 1^+} \left(\frac{1}{(x-5)(x+1)} + \ln(x-1) \right) = -\frac{1}{8} + \lim_{x \to 1^+} \ln(x-1) \overset{\lim_{x \to 1^+} \ln(x-1) = -\infty}{=} -\infty,$$

so $x = 1$ is a vertical asymptote. Note that the function is only defined for $x > 1$, and so $x = -1$ **is not a vertical asymptote!**

$$\lim_{x \to \infty} \left(\frac{1}{(x-5)(x+1)} + \ln(x-1) \right) = 0 + \lim_{x \to \infty} \ln(x-1) = \infty,$$

and so there are no horizontal asymptotes.

Example 2) Find the vertical and horizontal asymptotes for $y = \dfrac{3x^2}{(x-2)(x+1)}$.

Solution The vertical asymptotes are $x = 2$ and $x = -1$. For horizontal, consider:

$$\lim_{x \to \pm\infty} \frac{3x^2}{(x-2)(x+1)} \overset{\text{Divide top and bottom by } x^2}{=} \lim_{x \to \pm\infty} \frac{\left(\frac{3x^2}{x^2}\right)}{\left(\frac{(x-2)(x+1)}{x^2}\right)} = \lim_{x \to \pm\infty} \frac{3}{\left(1 - \frac{2}{x}\right)\left(1 + \frac{1}{x}\right)} = 3$$

Therefore, the horizontal asymptote is $x = 3$.

> **Note:** There can be more than one vertical asymptote. BUT because y is a FUNCTION, there can be only one horizontal asymptote (or none) as x approaches ∞ and one (or none) as x approaches $-\infty$.
>
> **Note:** YOU must decide whether you need to check the limits as x approaches $+\infty$ and $-\infty$ separately!

Slant Asymptotes

A **slant asymptote** is a straight line that a function "asymptotically" approaches as x approaches ∞ or $-\infty$ or both. All you have to do is check $\lim_{x \to \pm\infty} f(x)$.

Example 1) Find the slant asymptote(s) of the function $f(x) = 2x + \dfrac{1}{x+1}$.

Solution $\lim_{x \to \pm\infty} f(x)$

$= \lim_{x \to \pm\infty}\left(2x + \dfrac{1}{x+1}\right) \overset{\lim_{x \to \pm\infty}\left(\frac{1}{x+1}\right) = 0}{=} \lim_{x \to \pm\infty} 2x.$

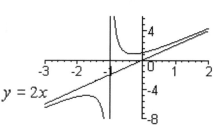

∴ The slant asymptote is $y = 2x$.

Just for interest, $x = -1$ is a vertical asymptote.

Example 2) Find the slant asymptotes, if any, of the function

$g(x) = \dfrac{1 - x + x^3}{3x^2 + 1} + e^x.$

Solution Remember that $\lim_{x \to \infty} e^x = \infty$ while $\lim_{x \to -\infty} e^x = 0$.

It would therefore be prudent to check $+\infty$ and $-\infty$ separately!

$\lim_{x \to \infty} g(x) = \lim_{x \to \infty}\left(\dfrac{1 - x + x^3}{3x^2 + 1} + e^x\right) \overset{\text{Divide top and bottom by } x^2.}{=} \lim_{x \to \infty}\left(\dfrac{\frac{1}{x^2} - \frac{1}{x} + x}{3 + \frac{1}{x^2}} + e^x\right) = \lim_{x \to \infty}\left(\dfrac{x}{3} + e^x\right)$

$\lim_{x \to -\infty} g(x) = \lim_{x \to -\infty}\left(\dfrac{1 - x + x^3}{3x^2 + 1} + e^x\right) \overset{\text{Divide top and bottom by } x^2.}{=} \lim_{x \to -\infty}\left(\dfrac{\frac{1}{x^2} - \frac{1}{x} + x}{3 + \frac{1}{x^2}} + e^x\right) \overset{\lim_{x \to -\infty} e^x = 0}{=} \lim_{x \to -\infty} \dfrac{x}{3}$

∴ There is a slant asymptote of $y = \dfrac{x}{3}$ as $x \to -\infty$ (but not $+\infty$!).

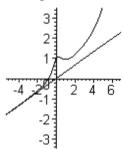

Intersection of Two Curves

When finding the intersection of $y = f(x)$ and $y = g(x)$ you

1) set $f(x) = g(x)$ and solve for x;
2) substitute each value of x into **either** $f(x)$ or $g(x)$ to find the corresponding y;
3) if it is fairly straightforward, sketch the graphs of the two curves so you have an idea of how many intersection points there are and approximately where to find them. **Use this as a guide but be prepared on occasion to be surprised if your guess doesn't tally with the math.**

Example 1) Find the intersection of the curves given by $y = f(x) = x^2$ and $y = g(x) = x + 6$.

Solution Set $x^2 = x + 6$. $\therefore x^2 - x - 6 = 0$. Factoring, $(x-3)(x+2) = 6$ and so $x = 3$ or $x = -2$. Since $f(3) = 9$ and $f(-2) = 4$, the intersection points are $(3, 9)$ and $(-2, 4)$.

Note 1: We often forget we want the intersection points and stop once we find the x values. **Go back to ONE of the original curves to find y.**

Note 2: It doesn't matter whether you go back to f or g to find the y values. For example, here $g(3) = 9$ and $g(-2) = 4$.

Example 2) Find the intersection of $y = \sin x$ and $y = \cos x$, for $0 \leq x \leq 2\pi$.

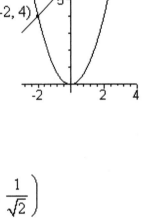

Solution Set $\sin x = \cos x$. The easiest way to tackle this one: divide both sides by $\cos x$, giving $\tan x = 1$. Now you don't need your calculator to solve this! In fact, if you do, your calculator would lead you astray. The calculator would give you, in radian mode, $\pi/4 \doteq 0.785$ radians. But **YOU** must realize that **tan is also positive in the third quadrant!** Don't ignore this since $0 \leq x \leq 2\pi$. The corresponding third quadrant angle is $5\pi/4$. And let's not forget: $\sin\left(\dfrac{\pi}{4}\right) = \dfrac{1}{\sqrt{2}}$ and $\sin\left(\dfrac{5\pi}{4}\right) = -\dfrac{1}{\sqrt{2}}$.

The interesection points are $\left(\dfrac{\pi}{4}, \dfrac{1}{\sqrt{2}}\right)$ and $\left(\dfrac{5\pi}{4}, -\dfrac{1}{\sqrt{2}}\right)$.

The Greatest Integer (or Floor) Function

The greatest integer function, also called the floor function, is usually denoted by $[[x]]$. **Every number** is either **an integer** or **lies between two integers, one above, one below**. The greatest integer function inputs a number.

If the number IS an integer, that's your answer: $[[4]] = 4$.

If not, your answer is the integer just below the number: $[[4.2]] = 4$.

Got it? Let's see.

Example 1) Evaluate each of the following:
(a) $[[1.7]]$ (b) $[[-2.3]]$ (c) $[[8]]$ (d) $[[-6]]$

Solution (a) 1   (b) -3 (c) 8   (d) -6

Example 2) Let $f(x) = [[x]]$, for $-2 \leq x \leq 3$.
Write $f(x)$ without using greatest integer notation and draw its graph.

Solution $f(x) = \begin{cases} -2, & \text{if } -2 \leq x < -1 \\ -1, & \text{if } -1 \leq x < 0 \\ 0, & \text{if } 0 \leq x < 1 \\ 1, & \text{if } 1 \leq x < 2 \\ 2, & \text{if } x = 2 \end{cases}$

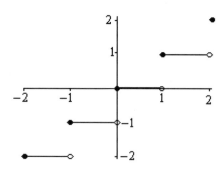

Example 3) Solve $[[2x+1]] = 6$.

Solution $[[2x+1]] = 6$ $\Leftrightarrow$ [2x+1 must be at least 6 but less than 7.] $6 \leq 2x+1 < 7 \Leftrightarrow 5 \leq 2x < 6 \Leftrightarrow 2.5 \leq x < 3$.

Two For You – Vertical and Horizontal Asymptotes

For the following, find the horizontal and vertical asymptotes:

1) $y = \dfrac{x^3}{(x-1)(x-2)}$ 2) $y = \dfrac{5x}{3x+4} + \dfrac{1}{e^x - 1}$

Answers 1) vertical asymptotes: $x = 1$, $x = 2$; no horizontal asymptote

2) vertical asymptotes: $x = -\dfrac{4}{3}$, $x = 0$

horizontal asymptotes: $y = \dfrac{5}{3}$ as $x \to \infty$ and $y = \dfrac{2}{3}$ as $x \to -\infty$

Two For You – Slant Asymptotes

Find the slant asymptotes, if any, for these functions:

1) $f(x) = \dfrac{-x^5 - 2x + 7}{2x - 5x^4}$ 2) $g(x) = \dfrac{e^{-x} + 4x^2}{x - 3}$

Answers 1) $y = \dfrac{1}{5}x$ as $x \to \pm\infty$ 2) $y = 4x$ as $x \to \infty$

Two For You – Intersection of Two Curves

Find the intersection of the two curves:

1) $y = x^3$ and $y = x^2 + x - 1$ 2) $y = \sin x$ and $y = -\cos x, -\pi \leq x \leq \pi$

Answers 1) $(1, 1)$ and $(-1, -1)$ 2) $\left(\dfrac{3\pi}{4}, \dfrac{1}{\sqrt{2}}\right)$ and $\left(-\dfrac{\pi}{4}, -\dfrac{1}{\sqrt{2}}\right)$

Two For You – The Greatest Integer (or Floor) Function

1) Evaluate each of the following: (a) $[[7]]$ (b) $[[-2.2]]$ (c) $[[-0.0001]]$

2)(a) Solve: $[[x^2]] = 1$ (Hint: $1 \leq x^2 < 2$)   (b) $[[x^3 - 1]] = -4$

Answers 1)(a) 7   (b) -3 (c) -1

2)(a) $-\sqrt{2} < x \leq -1$ or $1 \leq x < \sqrt{2}$ (b) $-3^{1/3} \leq x < -2^{1/3}$

Graphs with the Greatest Integer Function

The key here is to find the equations explicitly using the appropriate intervals.

Example 1) Let $f(x) = [[2x]]$, for $-1 \leq x \leq 1$.
Write $f(x)$ without using greatest integer notation
and draw its graph.

Solution $f(x) = \begin{cases} -2, & \text{if } -1 \leq x < -0.5 \\ -1, & \text{if } -0.5 \leq x < 0 \\ 0, & \text{if } 0 \leq x < 0.5 \\ 1, & \text{if } 0.5 \leq x < 1 \\ 2, & \text{if } x = 1 \end{cases}$

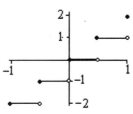

Example 2) Let $f(x) = x - [[x]]$, for $-2 \leq x \leq 2$.
Write $f(x)$ without using greatest integer notation
and draw its graph.

Solution $f(x) = \begin{cases} x+2, & \text{if } -2 \leq x < -1 \\ x+1, & \text{if } -1 \leq x < 0 \\ x, & \text{if } 0 \leq x < 0 \\ x-1, & \text{if } 1 \leq x < 2 \\ 0, & \text{if } x = 2 \end{cases}$

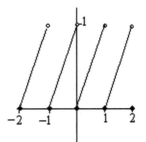

Example 3) Let $f(x) = [[x^2]]$, for $0 \leq x \leq 2$.
Write $f(x)$ without using greatest integer notation
and draw its graph.

Solution $f(x) = \begin{cases} 0, & \text{if } 0 \leq x < 1 \\ 1, & \text{if } 1 \leq x < \sqrt{2} \\ 2, & \text{if } \sqrt{2} \leq x < \sqrt{3} \\ 3, & \text{if } \sqrt{3} \leq x < 2 \\ 4, & \text{if } x = 2 \end{cases}$

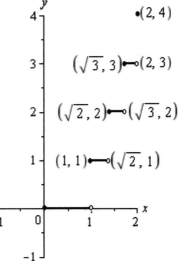

Properties of Exponents

Just as multiplication is a **short form for repeated addition** of the same number, exponentiation is a **short form for repeated multiplication** of the same number. For example, $5 \times 4 = 5 + 5 + 5 + 5$ and $5^4 = 5 \times 5 \times 5 \times 5$.

Here are the basic rules for exponents.

$$a^x a^y = a^{x+y} \qquad \frac{a^x}{a^y} = a^{x-y} \qquad (a^x)^y = a^{xy}$$

$$a^0 = 1 \qquad a^1 = a \qquad a^{-1} = \frac{1}{a}$$

$$a^{-x} = \frac{1}{a^x} \qquad \frac{1}{a^{-x}} = a^x$$

$$(ab)^x = a^x b^x \qquad \left(\frac{a}{b}\right)^x = \frac{a^x}{b^x}$$

Example 1) Evaluate each of the following:

(a) 4^3 (b) 0.1^0 (c) 5^{-1} (d) 0.1^{-3} (e) $\dfrac{12}{3^{-2}}$

Solution (a) 64   (b) 1   (c) $\dfrac{1}{5}$

(d) $0.1^{-3} = \left(\dfrac{1}{10}\right)^{-3} = 10^3 = 1000$ (e) $\dfrac{12}{3^{-2}} = 12 \cdot 9 = 108$

Example 2) Simplify each of the following:

(a) $x^3 x^7$ (b) $\dfrac{z^4 z^{-2}}{z^7}$ (c) $\left(\dfrac{2^{12} 3^4}{2^{-2} 3^{11}}\right)^{-2}$

Solution (a) $x^3 x^7 = x^{10}$ (b) $\dfrac{z^4 z^{-2}}{z^7} = z^{4-2-7} = z^{-5} = \dfrac{1}{z^5}$

(c) $\left(\dfrac{2^{12} 3^4}{2^{-2} 3^{11}}\right)^{-2} = \left(\dfrac{2^{14}}{3^7}\right)^{-2} = \dfrac{2^{-28}}{3^{-14}} = \dfrac{3^{14}}{2^{28}}$

Logarithms (Log Means "FIND THE EXPONENT!")

Logs cause headaches. I have to admit it. Students find logs hard. Why? I think in part it's because of the word "log". It seems to have no connection with what it represents in math. Then again, students seem to find exponents easy. So when you see "log", think, read, say out loud, "**FIND THE EXPONENT!**"

For example, $\log_2 16$: **Find the exponent** you need with base 2 to get a value of 16. 2 times 2 times 2 times 2 ... **FOUR!** You see, that's not so hard! Here are the rules.

$$\log_a(xy) = \log_a x + \log_a y \qquad \log_a\left(\frac{x}{y}\right) = \log_a x - \log_a y \qquad \log_a\left(x^y\right) = y\log_a x$$

$$\log_a 1 = 0 \qquad \log_a a = 1 \qquad \log_a a^{-1} = -1 \qquad \log_a\left(a^x\right) = x$$

DON'T confuse $\log_a(x^y)$ with $(\log_a x)^y$. For example,

$$\log_2(4^3) = \log_2\left(\left(2^2\right)^3\right) = \log_2(2^6) = 6$$

while $(\log_2 4)^3 = \log_2(2^2) \cdot \log_2(2^2) \cdot \log_2(2^2) = 2^3 = 8$.

Change of base formula: $\log_a x = \dfrac{\log_b x}{\log_b a}$ and in particular, $\log_a b = \dfrac{1}{\log_b a}$

Special bases: $\log x$ MEANS $\log_{10} x$ $\qquad$ $\ln x$ MEANS $\log_e x$

Example 1) Evaluate: (a) $\log 1000$ (b) $\log_2 \dfrac{1}{16}$ (c) $\log_3(27^4)$ (d) $(\log_3 27)^4$

Solution (a) $\log 1000 = 3$ (b) $\log_2 \dfrac{1}{16} = -4$ (c) $\log_3(27^4) = 4(3) = 12$

(d) $(\log_3 27)^4 = 3^4 = 81$

Example 2) Expand using log properties: (a) $\log_3\left(\dfrac{x^3 y^4}{z^5}\right)$ (b) $\log\left((x^2+1)(x^2-1)\right)$

Solution (a) $\log_3\left(\dfrac{x^3 y^4}{z^5}\right) = 3\log_3 x + 4\log_3 y - 5\log_3 z$

(b) $\log\left((x^2+1)(x^2-1)\right) = \log(x^2+1) + \log(x^2-1) \overset{\text{or}}{=} \log(x^2+1) + \log(x-1) + \log(x+1)$

Example 3) Change $\log_9 x$ to log base 4, log base 10, and log base e.

Solution $\log_9 x = \dfrac{\log_4 x}{\log_4 9} = \dfrac{\log x}{\log 9} = \dfrac{\ln x}{\ln 9}$

Basic Exponential Graphs

> Graphs with equations of the form $y = a^x$, where $a > 1$, are **really important**.

Example 1) Draw the graphs of $y = 2^x$ and $y = 3^x$ on the same axes.

Solution For bases $a > 1$:

1) $\lim\limits_{x \to \infty} a^x = \infty$ and the expression approaches ∞ **faster** as a increases.

2) $\lim\limits_{x \to -\infty} a^x = 0$ and the expression approaches 0 **faster** as a increases.

3) $\boxed{a^x \text{ IS ALWAYS} > 0}$

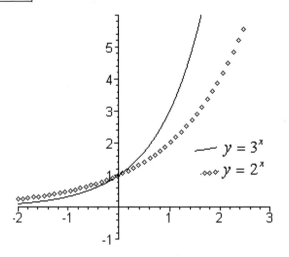

> Graphs with equations of the form $y = a^x$, where $0 < a < 1$, are **really really really almost as important**.

Example 2) Draw on the same axes the graphs of $y = \left(\dfrac{1}{2}\right)^x = 2^{-x}$ and $y = \left(\dfrac{1}{3}\right)^x = 3^{-x}$.

Solution

For bases $0 < a < 1$:

1) $\lim\limits_{x \to \infty} a^x = 0$ and the expression approaches 0 **faster** as a increases.

2) $\lim\limits_{x \to -\infty} a^x = \infty$ and the expression approaches ∞ **faster** as a increases.

3) $\boxed{a^x \text{ IS STILL ALWAYS} > 0}$

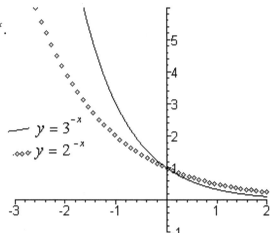

Two For You – Graphs with the Greatest Integer Function

1) Let $f(x) = 2x - [[x]]$, for $-2 \leq x \leq 1$.
Write $f(x)$ without using greatest integer notation and draw its graph.

2) Let $f(x) = [[x^2]]$, for $-2 \leq x \leq 0$. Write $f(x)$ without using $[[\ ]]$.

Answer 1) $f(x) = \begin{cases} 2x+2, & \text{if } -2 \leq x < -1 \\ 2x+1, & \text{if } -1 \leq x < 0 \\ 2x, & \text{if } 0 \leq x < 1 \\ 1, & \text{if } x = 1 \end{cases}$

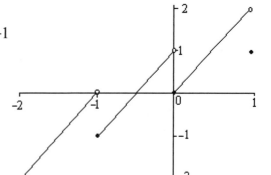

2) $f(x) = \begin{cases} 4, & \text{if } x = -2 \\ 3, & \text{if } -2 < x \leq -\sqrt{3} \\ 2, & \text{if } -\sqrt{3} < x \leq -\sqrt{2} \\ 1, & \text{if } -\sqrt{2} < x \leq -1 \\ 0, & \text{if } -1 < x \leq 0 \end{cases}$

Two For You – Properties of Exponents

1) Evaluate: (a) -7^0 (b) $(-3)^3$ (c) $\dfrac{1}{3^{-4}}$

2) Simplify: (a) $x^{1/2} x^5$ (b) $\left(\dfrac{x^6 y^7}{x^{-4} y^6}\right)^5$

Answers 1)(a) -1 (b) -27 (c) 81 2)(a) $x^{11/2}$ (b) $x^{50} y^5$

Two For You – Logarithms (Log Means "FIND THE EXPONENT!")

1)(a) Expand using properties of logs: $\log\left(a^{-2}b\sqrt{c}\right)^3$ (Hint: rewrite $\sqrt{c}$ as $c^{1/2}$.)

(b) Combine using properties of logs: $3\ln x - 4\ln y + 1$ (Hint: use $1 = \ln e$.)

2) Change $\log e$ (that is $\log_{10} e$) to base e.

Answers 1)(a) $-6\log a + 3\log b + \dfrac{3}{2}\log c$ (b) $\ln\left(\dfrac{ex^3}{y^4}\right)$ 2) $\log e = \dfrac{1}{\ln 10}$

Two For You – Basic Exponential Graphs

1)(a) As $x \to \infty$, which of these functions approaches ∞ faster: $y = e^x$ or $y = 10^x$?

(b) As $x \to -\infty$, which of these functions approaches 0 faster: $y = e^x$ or $y = 10^x$?

2)(a) As $x \to \infty$, which of these functions approaches faster: $y = e^{-x}$ or $y = 10^{-x}$?

(b) As $x \to -\infty$, which of these functions approaches ∞ faster: $y = e^{-x}$ or $y = 10^{-x}$?

Answers 1)(a) $y = 10^x$ (b) $y = 10^x$ 2)(a) $y = 10^{-x}$ (b) $y = 10^{-x}$

Basic Logarithm Graphs

> Graphs with equations of the form $y = \log_a x$, where $a > 1$, are **really important**.

Example 1) Draw the graphs of $y = \log_2 x$ and $y = \log_3 x$ on the same axes.

Solution For bases $a > 1$:

1) $\lim_{x \to \infty} \log_a x = \infty$ and the expression approaches ∞ **more slowly** as a increases.

2) $\lim_{x \to 0^+} \log_a x = -\infty$ and the expression approaches $-\infty$ **more slowly** as a increases.

3) To evaluate $\log_a x$, x MUST BE > 0

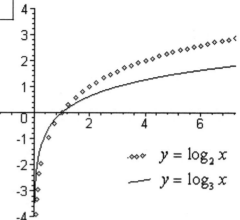

> Graphs with equations of the form $y = \log_a x$, where $0 < a < 1$, are **really almost as important**.

Example 2) Draw the graphs of $y = \log_{1/2} x$ and $y = \log_{1/3} x$ on the same axes.

Solution For bases $0 < a < 1$:

1) $\lim_{x \to 0^+} \log_a x = \infty$ and the expression approaches ∞ **more slowly** as a **decreases.**

2) $\lim_{x \to \infty} \log_a x = -\infty$ and the expression approaches $-\infty$ **more slowly** as a **decreases.**

3) To evaluate $\log_a x$, x MUST STILL BE > 0

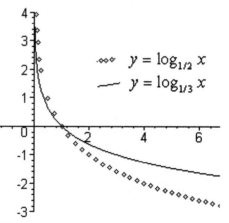

Inverse Formulas for Exponents and Logarithms

Remember: Log means "FIND THE EXPONENT!"

First, a review of the basics:

$$\log_a(xy) = \log_a x + \log_a y \qquad \log_a\left(\frac{x}{y}\right) = \log_a x - \log_a y \qquad \log_a(x^y) = y\log_a x$$

$$\log_a 1 = 0 \qquad \log_a a = 1 \qquad \log_a a^{-1} = -1 \qquad \log_a a^x = x$$

Change of base formula: $\log_a x = \dfrac{\log_b x}{\log_b a}$ and in particular, $\log_a b = \dfrac{1}{\log_b a}$

Special bases: $\log x$ MEANS $\log_{10} x$ $\quad$ $\ln x$ MEANS $\log_e x$. And now, the...

Inverse Formulas: $\quad a^{\log_a x} = x \quad 10^{\log x} = x \quad e^{\ln x} = x$
$\log_a(a^x) = x \quad \log(10^x) = x \quad \ln(e^x) = x$

Example 1) Simplify the following:

(a) $\log_3(3^{23})$ $\quad$ (b) $\log(10^{\sin x})$ $\quad$ (c) $\ln(e^{-2.37})$ $\quad$ (d) $\log_2(3^x)$

Solution (a) $\log_3(3^{23}) = 23$ $\quad$ (b) $\log(10^{\sin x}) = \sin x$ $\quad$ (c) $\ln(e^{-2.37}) = -2.37$
(d) $\log_2(3^x) = x\log_2 3$ We can't simplify further because **the bases don't match!**

Example 2) Simplify the following:

(a) $5^{\log_5 \pi}$ $\quad$ (b) $10^{\log(x+5)}$ $\quad$ (c) $e^{\ln(\ln x)}$ $\quad$ (d) $7^{\log_9 x}$

Solution (a) $5^{\log_5 \pi} = \pi$ $\quad$ (b) $10^{\log(x+5)} = x+5$ $\quad$ (c) $e^{\ln(\ln x)} = \ln x$
(d) $7^{\log_9 x}$ We can't make this simpler because **the bases don't match**!

Example 3) Simplify the following:

(a) $7\log_4(4^{2x+5})$ $\quad$ (b) $5\ln(e^{x^2})$ $\quad$ (c) $10^{3\log 2}$ $\quad$ (d) $e^{7\ln x}$ $\quad$ (e) $\log_2 4^x$

Solution (a) $7\log_4(4^{2x+5}) = 7(2x+5) = 14x+35$ $\quad$ (b) $5\ln(e^{x^2}) = 5x^2$
(c) $10^{3\log 2} = 10^{\log(2^3)} = 10^{\log 8} = 8$ $\quad$ (d) $e^{7\ln x} = e^{\ln(x^7)} = x^7$
(e) $\log_2 4^x = \log_2(2^2)^x = \log_2(2^{2x}) = 2x$

Solving Exponential Equations

To solve exponential and logarithm equations, you must be completely comfortable with:

$$a = b^x \Leftrightarrow x = \log_b a$$

b is the **base** x is the **exponent** a is, well, we will call it the **value**.

(Actually, a is the "power" but millions and millions of people confuse "power" with "exponent"! So, let's call it the value.)

Example 1) Find x in each of the following:

(a) $81 = 3^x$ (b) $5^{2x} = 12$ (c) $e^{x^2} = 9$ (d) $10^{4x+3} = 10^{2x-1}$ (e) $6^{3x+2} = -1$

Solution (a) $81 = 3^x \Leftrightarrow x = \log_3 81 = 4$

(b) $5^{2x} = 12 \Leftrightarrow 2x = \log_5 12$ $\underset{\text{optional, using the change of base formula}}{=} \dfrac{\ln 12}{\ln 5}$ [calculator] $\doteq 1.544$ and so $x \doteq 0.772$

(c) $e^{x^2} = 9 \Leftrightarrow x^2 = \ln 9 \Leftrightarrow x = \pm\sqrt{\ln 9}$ [optional] $\doteq \pm 1.482$

(d) $10^{4x+3} = 10^{2x-1} \Leftrightarrow 4x+3 = 2x-1 \Leftrightarrow 2x = -4 \Leftrightarrow x = -2$

(e) $6^{3x+2} = -1$ has no solution since, as long as a is positive, then $a^x > 0$ **ALWAYS!**

Example 2) Solve for x: $e^{2x} - 2e^x - 1 = 0$

Solution Treat this as a quadratic equation with variable e^x: $(e^x)^2 - 2e^x - 1 = 0$

Now use the quadratic formula for $aX^2 + bX + c = 0$ with $a = 1$, $b = -2$, $c = -1$, and $X = e^x$:

$e^x \underset{=}{\overset{\frac{-b \pm \sqrt{b^2 - 4ac}}{2a}}{}} \dfrac{-(-2) \pm \sqrt{(-2)^2 - 4(1)(-1)}}{2(1)} = \dfrac{2 \pm \sqrt{8}}{2} \underset{=}{\overset{\sqrt{8} = \sqrt{4 \cdot 2} = 2\sqrt{2}}{}} \dfrac{2 \pm 2\sqrt{2}}{2} = 1 \pm \sqrt{2}$

BUT $1 - \sqrt{2} < 0$ and $e^x > 0$!

$\therefore e^x = 1 + \sqrt{2}$ and so $x = \ln(1 + \sqrt{2}) \doteq 0.881$

Solving Logarithm Equations

To solve exponential and logarithm equations, you must be completely comfortable with:

$a = b^x \Leftrightarrow x = \log_b a$

b is the **base** x is the **exponent** a is, well, we will call it the **value**.

(Actually, a is the "power" but millions and millions of people confuse "power" with "exponent"! So, let's call it the value. Humour me!)

Example 1) Find x in each of the following:

(a) $\log_5(3x) = 2$ (b) $\ln(x^2) = -2$ (c) $\log_3(4x-1) = \log_3(6x-3)$

(d) $\ln(-1-x^2) = 1$

Solution (a) $\log_5(3x) = 2 \Leftrightarrow 3x = 5^2 = 25 \Leftrightarrow x = \dfrac{25}{3}$

(b) $\ln(x^2) = -2 \Leftrightarrow x^2 = e^{-2} = \dfrac{1}{e^2} \Leftrightarrow x = \pm\dfrac{1}{e}$

OR (b) $\ln(x^2) = -2 \Leftrightarrow$ [SINCE x CAN BE < 0, WE NEED TO INSERT "| |"!] $2\ln|x| = -2 \Leftrightarrow \ln|x| = -1 \Leftrightarrow x = \pm\dfrac{1}{e}$

(c) $\log_3(4x-1) = \log_3(6x-3) \Leftrightarrow 4x-1 = 6x-3 \Leftrightarrow 2 = 2x \Leftrightarrow x = 1$

(d) $\ln(-1-x^2) = 1 \Leftrightarrow e = -1-x^2$.

This has no solution since $e > 2$ while $-1-x^2 \leq -1$.

Example 2) Solve for x: $\log x + \log(2x-1) = \log 3$

Solution Note $x > 0$ and $x > \dfrac{1}{2}$ because of the domain of log functions!

So $x > \dfrac{1}{2}$ in this question. Using log properties:

$\log x + \log(2x-1) = \log 3 \Leftrightarrow \log(x(2x-1)) = \log 3 \Leftrightarrow 2x^2 - x = 3$

$\Leftrightarrow 2x^2 - x - 3 = 0 \Leftrightarrow (2x-3)(x+1) = 0 \Leftrightarrow x = \dfrac{3}{2}$ or $x = -1$

BUT $x > \dfrac{1}{2}$, and so $x = \dfrac{3}{2}$.

Two For You – Basic Logarithm Graphs

1)(a) As $x \to \infty$, which of these functions approaches ∞ faster:
$y = \ln x$ or $y = \log x$?

(b) As $x \to 0^+$, which of these functions approaches $-\infty$ faster:
$y = \ln x$ or $y = \log x$?

2)(a) As $x \to \infty$, which of these functions approaches $-\infty$ faster:
$y = \log_{1/e} x$ or $y = \log_{1/10} x$?

(b) As $x \to 0^+$, which of these functions approaches ∞ faster:
$y = \log_{1/e} x$ or $y = \log_{1/10} x$?

Answers 1)(a) $y = \ln x$ (b) $y = \ln x$ 2)(a) $y = \log_{1/e} x$ (b) $y = \log_{1/e} x$

Two For You – Inverse Formulas for Exponents and Logarithms

Simplify: 1)(a) $\log_7(7^{3x})$ (b) $\ln(e^{4x+e^x})$ (c) $\log_2(8^t)$

2)(a) $10^{\log 7}$ (b) $e^{4\ln(\sin x)}$ (c) $4^{\log_9 x}$

Answers 1)(a) $3x$ (b) $4x + e^x$ (c) $3t$
2)(a) 7 (b) $\sin^4(x)$ (c) No **easy** simplification since the bases are different.

Two For You – Solving Exponential Equations

1) Find x in each of the following:

(a) $2^x = 128$ (b) $7^{3x+1} = 49$ (c) $e^{e^x} = e$ (d) $10^{\sin x} = 1$ (e) $(-6)^{3x+6} = -216$

2) Solve for x: $10^{2x} + 2(10^x) - 1 = 0$

Answers 1)(a) $x = 7$ (b) $x = \dfrac{1}{3}$ (c) $x = 0$ (d) $x = k\pi, k \in \mathbb{Z}$ (e) $x = -1$

2) $x = \log(-1 + \sqrt{2})$ $(\because 10^x > 0 \therefore 10^x \neq -1 - \sqrt{2}.)$

Two For You – Solving Logarithm Equations

1) Find x in each of the following:

(a) $\log_5 x = 2$ (b) $\log_2(3x) = 4$ (c) $\ln(\ln x) = 0$ (d) $\log(3x) = \log(12 - x)$

2) Solve for x: $\ln(4x + 7) - \ln(x) = \ln 5$

Answers 1)(a) $x = 25$ (b) $\dfrac{16}{3}$ (c) $x = e$ (d) $x = 3$ 2) $x = 7$

The Derivative of $y = e^x$ and $y = a^x$

I **LOVE** this formula: $\dfrac{d(e^x)}{dx} = e^x$

e^x is its own derivative. At each x value, the slope equals the height.

But enough of my obsession. We have examples to do and two for you!

Example 1) Find $\dfrac{dy}{dx}$ for each of the following:

(a) $y = x^3 e^x$ (b) $y = e^{\sin x}$ (c) $y = e^{4x}$ (d) $y = e^{x \ln 4}$ (e) $y = 4^x$

Solution (a) $y = x^3 e^x$ $\therefore$ $\dfrac{dy}{dx}$ [Product Rule] $= x^3(e^x) + e^x(3x^2)$ [a little neater...] $= x^2 e^x(x+3)$

(b) $y = e^{\sin x}$ $\therefore$ $\dfrac{dy}{dx}$ [Chain Rule] $= e^{\sin x} \cos x$ (c) $y = e^{4x}$ $\therefore$ $\dfrac{dy}{dx}$ [Chain Rule] $= e^{4x}(4) = 4e^{4x}$

(d) $y = e^{x \ln 4}$ $\therefore$ $\dfrac{dy}{dx}$ [Chain Rule] $= e^{x \ln 4}(\ln 4)$ [Bring the "ln 4" to the left so this answer looks like the answer in (c). Remember ln 4 is just a constant.] $= (\ln 4)e^{(\ln 4)x}$

(e) $y = 4^x$. Remember a^x [$X = e^{\ln X}$. Here, $X = a^x$] $= e^{\ln(a^x)}$ [Log Property!] $= e^{x \ln a}$. So $y = 4^x$ [Here, $a^x = 4^x$] $= e^{\ln(4^x)} = e^{x \ln 4}$

$\therefore \dfrac{dy}{dx}$ [Now this is part (d) above!] $= e^{x \ln 4}(\ln 4)$ [Don't forget: $e^{x \ln 4} = 4^x$] $= 4^x \ln 4$ [for those who like their constants at the left...] $= (\ln 4)4^x$

Part (e) gives us the rule for the derivative of $y = a^x$: $\dfrac{d(a^x)}{dx} = a^x \ln a$.

Example 2) Find $\dfrac{dy}{dx}$: (a) $y = 2^x \tan x$ (b) $y = 5^{\sec x}$ (c) $y = \dfrac{3^{x^2}}{x}$

Solution (a) $y = 2^x \tan x$
$\therefore \dfrac{dy}{dx}$ [Product Rule] $= 2^x \sec^2 x + \tan x \, 2^x \ln 2$ [a little neater...] $= 2^x(\sec^2 x + \tan x \ln 2)$

(b) $y = 5^{\sec x}$ $\therefore \dfrac{dy}{dx}$ [Chain Rule] $= 5^{\sec x} \sec x \tan x \ln 5$

(c) $y = \dfrac{3^{x^2}}{x}$ $\therefore \dfrac{dy}{dx}$ [Quotient Rule] $= \dfrac{x 3^{x^2}(2x)\ln 3 - 3^{x^2}(1)}{x^2}$ [a little neater...] $= \dfrac{3^{x^2}(2x^2 \ln 3 - 1)}{x^2}$

The Derivative of $y = \ln x$ and $y = \log_a x$

My students are pretty comfortable with the formula $\dfrac{d(\ln x)}{dx} = \dfrac{1}{x}$. It is also true that $\dfrac{d(\ln|x|)}{dx} = \dfrac{1}{x}$. The second formula allows $x < 0$. The good news is that the derivative is the same!

Example 1) Given $\dfrac{d(\ln x)}{dx} = \dfrac{1}{x}$, prove that $\dfrac{d(\ln|x|)}{dx} = \dfrac{1}{x}$.

Solution Case 1) Let $x > 0$ so that $|x| = x$. $\therefore \dfrac{d(\ln|x|)}{dx} = \dfrac{d(\ln x)}{dx} = \dfrac{1}{x}$

Case 2) Let $x < 0$ so that $|x| = -x$. $\therefore \dfrac{d(\ln|x|)}{dx} = \dfrac{d(\ln(-x))}{dx} \overset{\text{Chain Rule!}}{=} \dfrac{-1}{(-x)} = \dfrac{1}{x}$

Example 2) Differentiate: (a) $y = \ln(3x)$ (b) $y = e^x \ln|1+3x|$ (c) $y = \ln(\ln x)$

Solution (a) $y = \ln(3x)$ $\therefore \dfrac{dy}{dx} \overset{\text{Chain Rule!}}{=} \dfrac{3}{3x} = \dfrac{1}{x}$ OR $\overset{\text{alternate method}}{y} \overset{\text{log property}}{=} \ln 3 + \ln x$ and so $\dfrac{dy}{dx} = \dfrac{1}{x}$

(b) $y = e^x \ln|1+3x|$ $\therefore \dfrac{dy}{dx} \overset{\text{Product Rule!}}{=} e^x \dfrac{3}{1+3x} + \ln|1+3x|(e^x) = e^x\left(\dfrac{3}{1+3x} + \ln|1+3x|\right)$

(c) $y = \ln(\ln x)$ $\therefore \dfrac{dy}{dx} \overset{\text{Chain Rule!}}{=} \dfrac{1}{\ln x}\left(\dfrac{1}{x}\right) \overset{\text{or}}{=} \dfrac{1}{x \ln x}$

Example 3) Find the derivative of $y = \log_2 x$.

Solution $y = \log_2 x \overset{\text{Change of Base formula}}{=} \dfrac{\log_e x}{\log_e 2} \overset{\log_e = \ln}{=} \dfrac{\ln x}{\ln 2} \overset{\text{Remember } \ln 2 \text{ is a constant.}}{=} \dfrac{1}{\ln 2}\ln x \therefore \dfrac{dy}{dx} = \dfrac{1}{\ln 2}\left(\dfrac{1}{x}\right) = \dfrac{1}{x \ln 2}$

Compare: $\dfrac{d(a^x)}{dx} = a^x \ln a$ $\dfrac{d(\log_a x)}{dx} = \dfrac{1}{x \ln a}$

Example 4) Differentiate: (a) $y = \log_5 |1+3x|$ (b) $y = \log_a(\log_a x)$

Solution (a) $y = \log_5 |1+3x|$ $\therefore \dfrac{dy}{dx} \overset{\text{Chain Rule!}}{=} \dfrac{3}{(1+3x)\ln 5}$

(b) $y = \log_a(\log_a x)$ [Be careful! Here, we make a change of base for the "outside" $\log_a$!] $= \dfrac{\ln(\log_a x)}{\ln a}$

$\therefore \dfrac{dy}{dx} \overset{\text{Chain Rule!}}{=} \dfrac{1}{(\log_a x)\ln a} \cdot \dfrac{1}{x \ln a} \overset{\text{much neater...}}{=} \dfrac{1}{x \log_a x (\ln a)^2}$

Logarithmic Differentiation Part I

Now this is a **fun** page. Really. Here we take TOUGH looking questions and make them easy. You see, logs aren't scary. They make tough questions easy.

Example 1) Find $\dfrac{dy}{dx}$ if $y = \ln\left(\dfrac{x^3(4x+5)^2}{(e^x+1)^5}\right)$.

Solution This looks very scary! Product rule, quotient rule, chain, log...

But... $y = \ln\left(\dfrac{x^3(4x+5)^2}{(e^x+1)^5}\right) \stackrel{\text{[Log Properties!]}}{=} 3\ln x + 2\ln(4x+5) - 5\ln(e^x+1)$

and so $\dfrac{dy}{dx} \stackrel{\text{[Don't forget the chain rule!]}}{=} \dfrac{3}{x} + \dfrac{8}{4x+5} - \dfrac{5e^x}{e^x+1}$. **EASY!**

Example 2) Slight complication...Find $\dfrac{dy}{dx}$ if $y = \dfrac{x^{1/3}\cos^3 x}{(x^4-x)^2}$.

Solution To use the log properties, we need to take "ln" of each side. We can only take ln of **positive numbers. So, first, take the absolute value of each side**, that is, set |Left Side| = |Right Side|. We should be concerned: does this change or restrict the original question? **NO!** Remember, $\dfrac{d\ln|x|}{dx} \stackrel{\text{[Say goodbye to absolute value!]}}{=} \dfrac{1}{x}$.

The absolute value disappears. We get the correct derivative with no restrictions!

$y = \dfrac{x^{1/3}\cos^3 x}{(x^4-x)^2}$ [Take the absolute value of each side.] $\therefore$ $|y| = \left|\dfrac{x^{1/3}\cos^3 x}{(x^4-x)^2}\right| = \dfrac{|x|^{1/3}|\cos x|^3}{|x^4-x|^2}$ and so

$\ln|y| = \ln\left(\dfrac{|x|^{1/3}|\cos x|^3}{|x^4-x|^2}\right) \stackrel{\text{[Log Properties!]}}{=} \dfrac{1}{3}\ln|x| + 3\ln|\cos x| - 2\ln|x^4-x|$

[Note: we differentiate **implicitly** on the left side.]

$\therefore \dfrac{1}{y}\dfrac{dy}{dx} = \dfrac{1}{3x} - \dfrac{3\sin x}{\cos x} - \dfrac{2(4x^3-1)}{x^4-x}$ and so

[Cross multiply by y. Use $\dfrac{3\sin x}{\cos x} = 3\tan x$.]

$\dfrac{dy}{dx} = y\left(\dfrac{1}{3x} - 3\tan x - \dfrac{2(4x^3-1)}{x^4-x}\right) \stackrel{\text{[Optional: replace y with the original expression.]}}{=} \dfrac{x^{1/3}\cos^3 x}{(x^4-x)^2}\left(\dfrac{1}{3x} - 3\tan x - \dfrac{2(4x^3-1)}{x^4-x}\right)$

Logarithmic Differentiation Part II: $\frac{d}{dx}\left(f(x)^{g(x)}\right)$

You can find $\frac{dy}{dx}$ if $y=(f(x))^3$: $\frac{dy}{dx}=3(f(x))^2 f'(x)$. This is "variable^{constant}".

You can find $\frac{dy}{dx}$ if $y=3^{f(x)}$: $\frac{dy}{dx}=3^{f(x)} f'(x)\ln 3$. This is "constant^{variable}".

But what about $\frac{dy}{dx}$ if $y=f(x)^{g(x)}$, that is, "variable^{variable}"?

Example 1) Differentiate $y=(2x+1)^{\sin x}$.

Solution Here we use the same method as Log Differentiation Part I. But because exponential functions are defined only when the base is positive...

> We allow the **arithmetic** expression $(-3)^3$ but not* the **function** $y=(-3)^x$!

...we don't need to take the absolute value of each side first.

$y=(2x+1)^{\sin x}$ ∴ $\ln y = \ln\left((2x+1)^{\sin x}\right)$ $\underset{\substack{\text{Use }\ln(a^b)=b\ln(a),\text{ with}\\a=2x+1\text{ and }b=\sin x.}}{=}$ $\sin x \ln(2x+1)$.

∴ $\frac{1}{y}\frac{dy}{dx} \underset{\text{Product Rule}}{=} \sin x\left(\frac{2}{2x+1}\right)+\ln(2x+1)(\cos x)$ and so

$\frac{dy}{dx}=y\left(\left(\frac{2\sin x}{2x+1}\right)+\ln(2x+1)(\cos x)\right) \underset{\substack{\text{Optional:}\\\text{replace }y.}}{=} (2x+1)^{\sin x}\left(\left(\frac{2\sin x}{2x+1}\right)+\ln(2x+1)(\cos x)\right)$

[Get a common denominator.]
$= (2x+1)^{\sin x}\left(\frac{2\sin x + (2x+1)\ln(2x+1)(\cos x)}{2x+1}\right)$

[Notice on the previous line that the exponential base is $2x+1$: the exponent on the top is $\sin x$ and on the bottom is 1.]
$= (2x+1)^{\sin x-1}\left(2\sin x + (2x+1)\ln(2x+1)(\cos x)\right)$

Example 2) Find $\frac{dy}{dx}$ if $y=(x^2+1)^{\cos x}$.

Solution $\ln y = \ln\left((x^2+1)^{\cos x}\right) = \cos x \ln(x^2+1)$ ∴ $\frac{1}{y}\frac{dy}{dx}=\cos x\left(\frac{2x}{x^2+1}\right)+\ln(x^2+1)(-\sin x)$

and so $\frac{dy}{dx}=y\left(\left(\frac{2x\cos x}{x^2+1}\right)-\sin x \ln(x^2+1)\right)=(x^2+1)^{\cos x}\left(\left(\frac{2x\cos x}{x^2+1}\right)-\sin x \ln(x^2+1)\right)$

[optional]
$= (x^2+1)^{\cos x-1}\left(2x\cos x - (x^2+1)\sin x \ln(x^2+1)\right)$

*Actually, we do in the area of mathematics called "complex analysis".

Two For You – The Derivative of $y = e^x$ and $y = a^x$

Differentiate: 1)(a) $y = e^{\cos x}$ (b) $y = \sqrt{e^{2x} + 4}$

2)(a) $y = 3^{\cos x}$ (b) $y = \sqrt{a^{2x} + 4}$, where $a > 0$

Answers 1)(a) $\dfrac{dy}{dx} = -e^{\cos x} \sin x$ (b) $\dfrac{dy}{dx} = \dfrac{e^{2x}}{\sqrt{e^{2x} + 4}}$

2)(a) $\dfrac{dy}{dx} = -3^{\cos x} (\ln 3) \sin x$ (b) $\dfrac{dy}{dx} = \dfrac{a^{2x} \ln a}{\sqrt{a^{2x} + 4}}$

Two For You – The Derivative of $y = \ln x$ and $y = \log_a x$

Differentiate: 1) $y = \ln|\cos x| + \ln|\sec x + \tan x|$ 2) $y = \log(\log(x))$

Answers 1) $-\tan x + \sec x$ 2) $\dfrac{1}{x \log x (\ln 10)^2}$

Two For You – Logarithmic Differentiation Part I

Find $\dfrac{dy}{dx}$ in the following: 1) $y = \ln\left(\dfrac{x^5 \ln x}{\tan x}\right)$ 2) $y = \left(\dfrac{(5x-1)\sin^5 x}{e^x + 5}\right)^3$

Answers 1) $\dfrac{5}{x} + \dfrac{1}{x \ln x} - \dfrac{\sec^2 x}{\tan x}$ 2) $3y\left(\dfrac{5}{5x-1} + 5\cot x - \dfrac{e^x}{e^x + 5}\right)$

Two For You – Logarithmic Differentiation Part II: $\dfrac{d}{dx}\left(f(x)^{g(x)}\right)$

Differentiate: 1) $y = x^{5x+1}$ 2) $y = (\tan x)^{\ln x}$

Answers 1) $\dfrac{dy}{dx} = x^{5x+1}\left(\dfrac{5x+1}{x} + 5\ln x\right) \stackrel{\boxed{\text{optional}}}{=} x^{5x}(5x + 1 + 5x\ln x)$

2) $\dfrac{dy}{dx} = (\tan x)^{\ln x}\left(\dfrac{\ln x \sec^2 x}{\tan x} + \dfrac{\ln(\tan x)}{x}\right) = (\tan x)^{\ln x - 1}\left(\dfrac{(x \ln x \sec^2 x) + \tan x \ln(\tan x)}{x}\right)$

Integrals Yielding ln: $\int \frac{\left(\frac{du}{dx}\right)}{u} dx = \ln|u| + C$

Compare these three integrals:

1) $\int \frac{1}{x^2+1} dx$ 2) $\int \frac{2x}{(x^2+1)^2} dx$ 3) $\int \frac{2x}{x^2+1} dx$.

Which one fits into the pattern $\int \frac{\left(\frac{du}{dx}\right)}{u} dx$? In all three examples, $u = x^2 + 1$.

Integral 1) misses the pattern because $\frac{du}{dx} = 2x$ is nowhere to be found.

Integral 2) misses the pattern because the exponent on u in the bottom is 2.

Integral 3) is like Goldilock's choice of porridge: just right!

In fact, $\int \frac{1}{x^2+1} dx = \arctan x + C$, which you may not have yet studied!

In fact, $\int \frac{2x}{(x^2+1)^2} dx$ is the Chain Rule in Reverse with the Power Rule: $\int u^{-2} \frac{du}{dx} dx$:

$$\int \frac{2x}{(x^2+1)^2} dx = \int (x^2+1)^{-2} 2x\, dx = \frac{(x^2+1)^{-1}}{-1} + C = \frac{-1}{x^2+1} + C$$

But $\int \frac{2x}{x^2+1} dx = \ln|x^2+1| + C$ $\boxed{x^2+1 > 0 \text{ so we don't need absolute value.}}$ $= \ln(x^2+1) + C$

Example 1) Evaluate the following integrals:

(a) $\int \frac{1+\sin x}{x - \cos x} dx$ (b) $\int \frac{e^{3x}}{e^{3x} - 5} dx$ (c) $\int \frac{1}{x \ln x} dx$

Solution (a) $\int \frac{1+\sin x}{x - \cos x} dx = \ln|x - \cos x| + C$

(b) $\int \frac{e^{3x}}{e^{3x} - 5} dx$ $\boxed{\text{Adjust the multiplicative constant.}}$ $= \frac{1}{3} \int \frac{3e^{3x}}{e^{3x} - 5} dx = \frac{1}{3} \ln|e^{3x} - 5| + C$

(c) $\int \frac{1}{x \ln x} dx$ $\boxed{\text{Here, } u = \ln x \text{ and } \frac{du}{dx} = \frac{1}{x}.}$ $= \int \frac{\left(\frac{1}{x}\right)}{\ln x} dx = \ln|\ln x| + C$

A Degree of Knowledge about Angles

Angle Facts! Remember that, in degree measure, we always include the degree symbol. In radian measure, we sometimes (your choice) omit "radians".

Triangles: The sum of the angles in a triangle is $180° = \pi$ radians.
If all angles are equal, each **interior** angle is $60° = \dfrac{\pi}{3}$. Also, if the angles are equal, so are the sides. (In a triangle: **equiangular $\Leftrightarrow$ equilateral**)

Quadrilaterals: The sum of the angles in a quadrilateral is $360° = 2\pi$ radians.
If the quadrilateral is a square (or rectangle), each **interior** angle is $90° = \dfrac{\pi}{2}$.
(In a quadrilateral: **equiangular $\not\Leftrightarrow$ equilateral**)

Polygons: The sum of the angles in a polygon with n sides is

[In Degree Measure!] $\quad 180(n-2)° = 180n° - 360°\qquad$ [In Radian Measure!] $\quad(n-2)\pi$ radians $= n\pi - 2\pi$ radians

If this is a "**regular**" polygon (**both equiangular and equilateral**), each **interior** angle is
$\dfrac{(n-2)180°}{n} = 180° - \dfrac{360°}{n} = \pi - \dfrac{2\pi}{n}$ radians. So, for example, if $n = 4$, then
the interior angle $= 180° - \dfrac{360°}{4} = 90° = \dfrac{\pi}{2}$ and this is what we found for rectangles.

Parallel Lines: Given $\;l_1 \parallel l_2\;$ [|| is the symbol for parallel.] :

Alternate Angles: $a = c$ and $b = d$ **(Z rule)**

Corresponding Angles: $a = e$ and $d = f$ **(F Rule)**

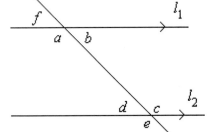

Interior Angles (on the same side of the transversal):
$a + d = b + c = 180° = \pi$ radians **(C Rule)**

Circles: If d is the angle at the centre of the first circle, then $d = 2e$.
In the second (*funny looking*) circle:
$a = b = (180 - c)° \;$ [or, if we are using radian measure,] $\;= \pi - c$

(a and b are said to be **subtended by the same arc**.)

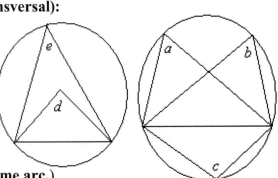

The Pythagorean Theorem

Pythagoras and friends proved this famous theorem around 500 BC, although it was known long before then. In fact, it caused Pythagoras some mathematical heartburn. He believed that **all** numbers were either whole numbers or rational numbers (quotients of whole numbers.) Then he took a closer look at the $(45°, 45°, 90°)$ triangle, with two equal sides of length 1. The hypotenuse turned out to have length $\sqrt{2}$ and a little mathematical ingenuity shows that $\sqrt{2}$ is NOT a rational number, let alone a whole number. The embarrassing truth is that the "Pythagorean Society" tried to hide the existence of this fact.

> **The Pythagorean Theorem**
> Let c be the hypotenuse of a right triangle whose other two sides are a and b.
> Then $a^2 + b^2 = c^2$.

$a^2 + b^2 = c^2$

The **converse** is also true: If a triangle with sides a, b, and c satisfies $a^2 + b^2 = c^2$, then the angle opposite side c is $90°$.

Example 1) In the diagram, find x and then find y.

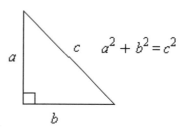

Solution $x^2 = 3^2 + 4^2 = 25$ and so $x = 5$ [x is a length and so $x \geq 0$.]

$y^2 = 7^2 - 5^2 = 24$ and so $y = \sqrt{24}$ [Back of Book!] $= 2\sqrt{6}$

Example 2) Prove that $\triangle ABC$ is a right triangle. Which vertex has the right angle?

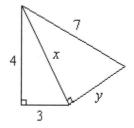

Solution Since $AB^2 + AC^2 = 13 + 6 = 19 = BC^2$, therefore, by the converse of the Pythagorean Theorem, $\triangle ABC$ is a right triangle, with $\angle A = 90°$.

Note : The Cosine Law states that in a triangle with sides a, b, and c, $c^2 = a^2 + b^2 - 2ab\cos(\theta)$, where θ is the angle between a and b. If $\theta = 90°$, then the cosine law becomes the Pythagorean Theorem!

Similar Triangles

Two triangles are similar if all three pairs of corresponding angles are equal and corresponding sides are in the same ratio. In this pair of similar triangles............ →

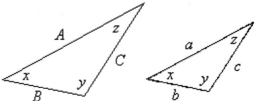

$\dfrac{A}{a} = \dfrac{B}{b} = \dfrac{C}{c}$. Note that side "$A$" in the big triangle and side "a" in the small triangle correspond because they are both between angles x and z.

Conditions for similarity: We can conclude two triangles are similar if we know

SSS: all three pairs of corresponding sides are in the same ratio **or**

SAS: two pairs of corresponding sides are in the same ratio and the contained angles are equal **or**

AAA: all three pairs of corresponding angles are equal **or**

AA: if two pairs of angles are equal, the third pair must be as well, so the triangles are similar by AAA.

Example 1) In the diagram at the top, if $A = 10$, $B = 8$, and $a = 4$, find C, b, and c.

Solution $\dfrac{A}{a} = \dfrac{10}{4} = \dfrac{5}{2}$. Therefore, $\dfrac{B}{b} = \dfrac{8}{b} = \dfrac{5}{2}$ and so $b = \dfrac{16}{5}$. As for c and C, we know that $\dfrac{C}{c} = \dfrac{5}{2}$ and so $C = \dfrac{5c}{2}$. However, without knowing either C or c (or at least one of the angles), we can not find either the value of C or c! (Don't you hate trick questions?!)

Example 2) In the picture to the right, find a formula for s in terms of d.

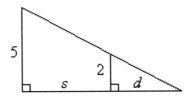

Solution By similar triangles (AAA), $\dfrac{s+d}{d} = \dfrac{5}{2}$. So, $2s + 2d = 5d$ and $s = \dfrac{3d}{2}$.

Two For You – Integrals Yielding ln: $\int \frac{\left(\frac{du}{dx}\right)}{u} dx = \ln|u| + C$

Evaluate the following integrals:

1) $\int \frac{x^2}{2x^3 + 1} dx$

2) $\int \frac{1}{x \ln(x) \ln(\ln(x))} dx$ (Hint: let $u = \ln(\ln(x))$.)

Answers 1) $\frac{1}{6} \ln|2x^3 + 1| + C$ 2) $\ln|\ln(\ln x)| + C$

Two For You – A Degree of Knowledge about Angles

1)(a) In the triangle, find x in radian measure.

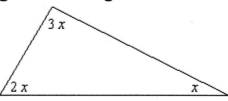

(b) In the circle, find y, z, and w. Assume the angle labeled z is at the centre of the circle.

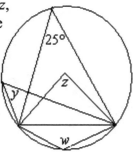

2)(a) What is the interior angle, in degrees, of a regular hexagon? (Hint: hex = six)

(b) What is the interior angle, in degrees, of a regular 180-gon? (Hint: $n = 180$)

Answers 1)(a) $\frac{\pi}{6}$ (b) $y = 25°$ $z = 50°$ $w = 155°$ 2)(a) $120°$ (b) $178°$

Two For You – The Pythagorean Theorem

1) A Pythagorean Triple is an ordered triple of numbers (a,b,c) satisfying $a^2+b^2=c^2$. Complete these Pythagorean Triples:

(a) $(5,12,c)$ (b) $(\sqrt{5},2\sqrt{3},c)$ (c) $(1,b,2)$ (d) $(a,8,7)$

2) Two friends say goodbye. One drives north averaging 100 km/h. The other drives west at an average speed of 80 km/h. How far apart are they after two hours?

Answers 1)(a) $c=13$ (b) $c=\sqrt{17}$ (c) $b=\sqrt{3}$
(d) No solution since c must be greater than each of a and b!
2) $40\sqrt{41}$ km or about 256.1 km

Two For You – Similar Triangles

1) In the triangles at the right, suppose $A = 7$, $B = 4$, $b = 2$, and $c = 3$. Find a, and C.

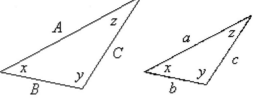

2) Use the similar triangles below to find an expression for a in terms of b.

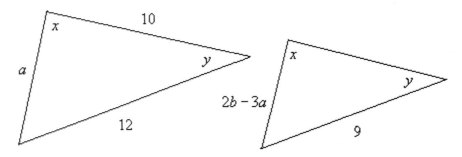

Answers 1) $a = 7/2$ and $C = 6$ 2) $a = 8b/15$

Radian Measure of an Angle

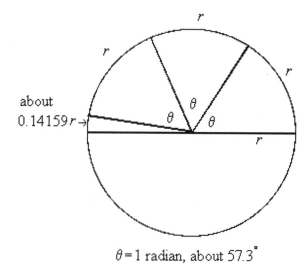

$\theta = 1$ radian, about $57.3°$

For any circle, take the circle's radius r. Wrap r around the circumference. This will create an angle just a little over 57 degrees. We call this 1 radian. If you wrap another radius around the circle starting at the end of the first wrapped r, and then another, and then another, you will find that the semi-circle will have 3 **and about** 0.14159 radii wrapped around it in all. We give the number of radii wrapped around the semi-circle the name π. So

$\pi \doteq 3.14159$ and π radians $= 180°$.

It is customary to omit the word "radians" but not "degrees"!

Example 1) Convert the following radian measures to degrees:

(a) $\dfrac{\pi}{2}$ (b) $\dfrac{7\pi}{3}$ (c) 2

Solution $\because \pi$ radians $= 180°$ $\therefore$ 1 radian $= \left(\dfrac{180}{\pi}\right)°$. From now on, we will omit "radians"

(a) $\dfrac{\pi}{2} = \dfrac{\pi}{2}\left(\dfrac{180}{\pi}\right)° = 90°$ (b) $\dfrac{7\pi}{3} = \dfrac{7\pi}{3}\left(\dfrac{180}{\pi}\right)° = 420°$

(c) $2 = 2\left(\dfrac{180}{\pi}\right)° = \left(\dfrac{360}{\pi}\right)° \doteq 2(57.3)° = 114.6°$

Example 2) Convert the following degree measures to radians:

(a) $45°$ (b) $-40°$ (c) $\pi°$

Solution Since $180° = \pi$ radians, $\therefore$ 1 degree $= \left(\dfrac{\pi}{180}\right)$ radians.

(a) $45° = 45\left(\dfrac{\pi}{180}\right) = \dfrac{\pi}{4}$ (b) $-40° = -40\left(\dfrac{\pi}{180}\right) = -\dfrac{2\pi}{9}$

(c) $\pi° = \pi\left(\dfrac{\pi}{180}\right) = \dfrac{\pi^2}{180} \doteq 0.055$

Basic Trigonometric Ratios: SOH CAH TOA

SOH CAH TOA gives us an easy way to remember how to define the sine, cosine and tangent of an angle in a right triangle, that is, a triangle with a 90° angle. Let's start with the basics.

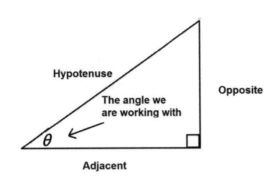

Label the sides of a right triangle based on the angle θ you are working with.

H – Hypotenuse: The side across from (that is, not touching) the right angle. It is the longest side of the triangle.

O – Opposite: The side across from θ.

A – Adjacent: The side that forms θ along with the hypotenuse.

Now you can easily remember the basic trigonometric ratios using the acronym **SOH CAH TOA**.

SOH: $\sin(\theta) = \dfrac{O}{H} = \dfrac{Opposite}{Hypotenuse}$

CAH: $\cos(\theta) = \dfrac{A}{H} = \dfrac{Adjacent}{Hypotenuse}$

TOA: $\tan(\theta) = \dfrac{O}{A} = \dfrac{Opposite}{Adjacent}$

Example 1) State the sine, cosine and tangent ratios for the given angle in the triangle below. Do not reduce your fractions.

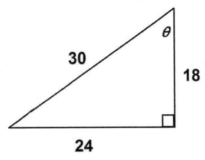

Solution

$\sin(\theta) = \dfrac{O}{H} = \dfrac{24}{30}$

$\cos(\theta) = \dfrac{A}{H} = \dfrac{18}{30}$

$\tan(\theta) = \dfrac{O}{A} = \dfrac{24}{18}$

Using SOH CAH TOA to Find Missing Sides and Angles

There are lots of ways to find information in triangles: the angles add to 180°; the Pythagorean Theorem; the sine and cosine laws; similar triangles, and more.

SOH CAH TOA gives us an easy way to find missing side lengths or angles in a **right triangle**.

Example 1) Find the length of the side labeled x.

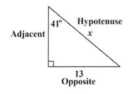

Solution

Label the sides of the triangle opposite, adjacent and hypotenuse based on the 41° angle as show as shown.
We want H – Hypotenuse. We know the angle and O – Opposite.

This means we should use SOH, that is sine.

$$\sin(\theta) = \frac{O}{H}$$

$$\sin(41°) = \frac{13}{x}$$

$$\frac{\sin(41°)}{1} = \frac{13}{x}$$

<u>Cross multiply.</u>

$$x\sin(41°) = 13 \cdot 1$$

$$x = \frac{13}{\sin(41°)}$$

<u>Make sure your calculator is in degree mode.</u>

$$x \doteq 19.8$$

Example 2) Find the angle θ. Give your answer in degrees rounded to one decimal.

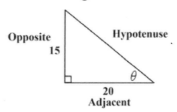

Solution

Label the sides of the triangle opposite, adjacent and hypotenuse based on the angle θ, as shown.

We have O – opposite and A – adjacent.

This means we should use TOA, that is tangent.

$$\tan(\theta) = \frac{O}{A}$$

$$\tan(\theta) = \frac{15}{20} = 0.75$$

$$\therefore \theta = \tan^{-1}(.75)\,*$$

<u>Make sure your calculator is in degree mode.</u>

$$\theta \doteq 36.9°$$

* To evaluate $\tan^{-1}$ on most calculators, you press the second function (or inverse) button and then the tan button.

Angles in Standard Position

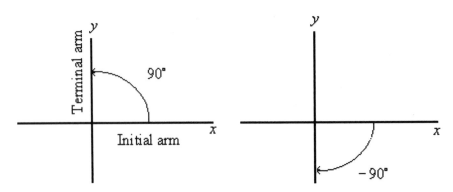

POSITIVE angles are drawn **COUNTER-CLOCKWISE** from the positive x axis.

NEGATIVE angles are drawn **CLOCKWISE** from the positive x axis.

Example 1)(a) Give, in degree measure, **all** the angles which have the same initial and terminal arms as (i) $90°$ (ii) $-90°$, and

(b) draw both $450°$ and $-450°$ in standard position.

Solution (a)(i) Let k be a **positive** integer. Then all the angles $90°$, $450°$, $810°$, and in general, $(90 + 360k)°$ have the same initial and terminal arms. Also, $90°$, $-270°$, $-630°$, and in general, $(90 - 360k)°$ have the same initial and terminal arms.

(ii) $-90°$, $-450°$, and in general, $(-90 - 360k)°$ have the same initial and terminal arms. Also, $-90°$, $270°$, and in general, $(-90 + 360k)°$ have the same initial and terminal arms.

SUMMARY

(i) $(90 + 360k)°$, for **any integer** k, has the same initial and terminal arms as $90°$.

(ii) $(-90 + 360k)°$, for **any integer** k, has the same initial and terminal arms as $-90°$.

(b)

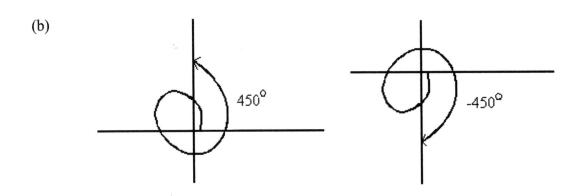

Two For You – Radian Measure of an Angle

1) Convert to degree measure: (a) $\pi/6$ (b) -1.8

2) Convert to radian measure: (a) $-60°$ (b) $12°$

Answers 1)(a) $30°$ (b) $\left(\dfrac{-324}{\pi}\right)° \doteq -103.1°$ 2)(a) $-\dfrac{\pi}{3}$ (b) $\dfrac{12\pi}{180} = \dfrac{\pi}{15} \doteq 0.21$

Two For You – Basic Trigonometric Ratios: SOH CAH TOA

1) Label the sides of the triangle to the right opposite, adjacent and hypotenuse based on
   (i) the $37°$ angle   (ii) the $53°$ angle

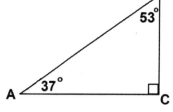

2) State the sine, cosine and tangent ratios for the angle θ in the triangle to the right.

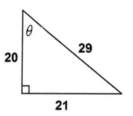

Answers 1)(i) side AB = Hypotenuse, side BC = Opposite, side AC = Adjacent
(ii) side AB = Hypotenuse, side AC = Opposite, side BC = Adjacent

2) $\sin(\theta) = \dfrac{21}{29}$ $\cos(\theta) = \dfrac{20}{29}$ $\tan(\theta) = \dfrac{21}{20}$

Two For You – Using SOH CAH TOA to Find Sides and Angles

1) Find the side labeled x. (Hint: Use $\tan(30°) = \frac{1}{\sqrt{3}}$.)

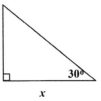

2) Find the angle θ in two different ways. Give your answer in degrees rounded to one decimal.

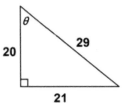

Answers 1) $x = 15\sqrt{3} \doteq 26$

2) $\theta \doteq 46.4°$. You can find θ using a sine ratio, cosine ratio or tangent ratio.

Two For You – Angles in Standard Position

1) Give all angles in standard position having the same terminal arm as

(a) $45°$ (b) $-60°$.

2) Redo question 1, giving your answers in radian measure.

Answers 1)(a) $(45 + 360k)°$, for $k \in \mathbb{Z}$ (b) $(-60 + 360k)°$, for $k \in \mathbb{Z}$

2)(a) $\frac{\pi}{4} + 2k\pi$, for $k \in \mathbb{Z}$ (b) $-\frac{\pi}{3} + 2k\pi$, for $k \in \mathbb{Z}$

Related Angles in Standard Position

You will find this topic **REALLY USEFUL** when you study Trigonometric Functions!

Here, we will use the following convention for angles:

Quadrant	Angle	
1	Degrees: $0° < \theta < 90°$	or Radians: $0 < \theta < \dfrac{\pi}{2}$
2	Degrees: $90° < \theta < 180°$	or Radians: $\dfrac{\pi}{2} < \theta < \pi$
3	Degrees: $180° < \theta < 270°$	or Radians: $\pi < \theta < \dfrac{3\pi}{2}$
4	Degrees: $-90° < \theta < 0°$	or Radians: $-\dfrac{\pi}{2} < \theta < 0$

Let $0 < \theta < 90°$. Draw θ in standard position and let it puncture the unit circle at $P(x, y)$. Drop a perpendicular from P to the x axis. Now you have a right triangle with hypotenuse of length 1. "Triangle Trig" gives $\cos(\theta) = x$, $\sin(\theta) = y$, and $\tan(\theta) = y/x$. Now draw three congruent triangles, one in each of the other three quadrants.

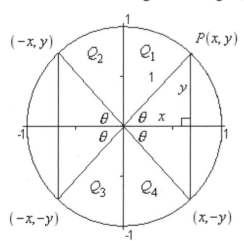

Each triangle has the angle θ at the origin.

The angle in STANDARD POSITION in quadrant TWO (Q2) that relates to θ is $180° - \theta$.

The angle in STANDARD POSITION in quadrant THREE (Q3) that relates to θ is $180° + \theta$.

The angle in STANDARD POSITION in quadrant FOUR (Q4) that relates to θ is $-\theta$.

Example 1) Find the related angles in each of the other three quadrants for

(a) $40°$ (b) $200°$ (c) $\dfrac{7\pi}{9}$ radians

Solution If θ is a first quadrant angle, use $180° - \theta$, $180° + \theta$, and $-\theta$ or the radian equivalents.

II: If the given angle is not in the first quadrant, first find the first quadrant relative, call this θ, and go back to Solution I!

(a) Quad 2: $180° - 40° = 140°$ Quad 3: $180° + 40° = 220°$ Quad 4: $-40°$

(b) Quad 1: $200° - 180° = 20°$ Quad 2: $180° - 20° = 160°$ Quad 4: $-20°$

(c) Quad 1: $\pi - \dfrac{7\pi}{9} = \dfrac{2\pi}{9}$ Quad 3: $\pi + \dfrac{2\pi}{9} = \dfrac{11\pi}{9}$ Quad 4: $-\dfrac{2\pi}{9}$

Trig Ratios for the $(30°, 60°, 90°) \equiv (\pi/6, \pi/3, \pi/2)$ Triangle

You DON'T have to memorize these ratios. It is so easy to take 30 seconds and redevelop them!

Draw an **equilateral** (and therefore **equiangular**) triangle. Each angle is $60° = \dfrac{\pi}{3}$ radians. Drop a perpendicular from one vertex. This divides the triangle into two congruent triangles, with angles of $30°$, $60°$, and $90°$ or, in radians, $\dfrac{\pi}{6}$, $\dfrac{\pi}{3}$, and $\dfrac{\pi}{2}$. Let each side in the original triangle have length 2. Then in each of the two congruent triangles, the sides (**using congruency and Pythagoras!**) are $2, 1,$ and $\sqrt{3}$.

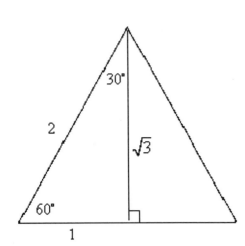

For $30°$, 1 is **opposite** and $\sqrt{3}$ is **adjacent**.
For $60°$, $\sqrt{3}$ is **opposite** and 1 is **adjacent**.
For both, 2 is the **hypotenuse**.

SOHCAHTOA!
SIN = O/H
COS = A/H
TAN = O/A
CSC = H/O
SEC = H/A
COT = A/O

(**Important: Do this construction yourself at least THREE TIMES.**)

Example 1) State the sine, cosine and tangent ratios for $30°$ and $60°$.

Solution

$$\sin(30°) = \sin\left(\frac{\pi}{6}\right) = \frac{1}{2} \qquad \cos(30°) = \cos\left(\frac{\pi}{6}\right) = \frac{\sqrt{3}}{2} \qquad \tan(30°) = \tan\left(\frac{\pi}{6}\right) = \frac{1}{\sqrt{3}}$$

$$\sin(60°) = \sin\left(\frac{\pi}{3}\right) = \frac{\sqrt{3}}{2} \qquad \cos(60°) = \cos\left(\frac{\pi}{3}\right) = \frac{1}{2} \qquad \tan(60°) = \tan\left(\frac{\pi}{3}\right) = \sqrt{3}$$

Trig Ratios for the $(45°, 45°, 90°) \equiv (\pi/4, \pi/4, \pi/2)$ Triangle

You DON'T have to memorize these ratios. It is so easy to take 30 seconds and redevelop them!

Draw a right isosceles triangle, that is, draw a $90°$ angle and make the two attached sides equal. Therefore, the triangle has angles of $45°$, $45°$, and $90°$ or in radians, $\frac{\pi}{4}$, $\frac{\pi}{4}$, and $\frac{\pi}{2}$. Let the two equal sides have length 1. Then (**using Pythagoras!**), the hypotenuse has length $\sqrt{2}$.

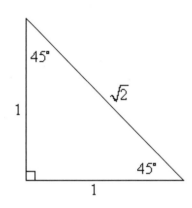

For $45°$, 1 is **opposite** and 1 is **adjacent**. The **hypotenuse** is $\sqrt{2}$.

SOHCAHTOA!
SIN = O/H
COS = A/H
TAN = O/A
CSC = H/O
SEC = H/A
COT = A/O

Example 1) State the sine, cosine and tangent ratios for $45°$.

Solution $\sin(45°) = \sin\left(\frac{\pi}{4}\right) = \frac{1}{\sqrt{2}}$ $\quad \cos(45°) = \cos\left(\frac{\pi}{4}\right) = \frac{1}{\sqrt{2}} \quad \tan(45°) = \tan\left(\frac{\pi}{4}\right) = 1$

Example 2) Given that $\sin^2(\theta) = \frac{1-\cos(2\theta)}{2}$, find the exact value of $\sin\left(\frac{\pi}{8}\right)$.

Solution Taking $\theta = \frac{\pi}{8}$, we have $\sin^2\left(\frac{\pi}{8}\right) = \frac{1-\cos\left(\frac{\pi}{4}\right)}{2} = \frac{1-\frac{1}{\sqrt{2}}}{2}$

[Get a common denominator **ON THE TOP** and simplify.] $= \frac{\sqrt{2}-1}{2\sqrt{2}}$ [Some of us love to rationalize the bottom!] $= \frac{2-\sqrt{2}}{4}$

and so $\sin\left(\frac{\pi}{8}\right)$ [Take the **POSITIVE** root since the angle is in the first quadrant.] $= \sqrt{\frac{2-\sqrt{2}}{4}} \doteq 0.383.$

Trig Ratios for 30°, 45°, 60° (and More)—A Table

degrees	radians	sin	cos	tan	csc	sec	cot
0	0	0	1	0	undefined	1	undefined
30	$\dfrac{\pi}{6}$	$\dfrac{1}{2}$	$\dfrac{\sqrt{3}}{2}$	$\dfrac{1}{\sqrt{3}}$	2	$\dfrac{2}{\sqrt{3}}$	$\sqrt{3}$
45	$\dfrac{\pi}{4}$	$\dfrac{1}{\sqrt{2}}$	$\dfrac{1}{\sqrt{2}}$	1	$\sqrt{2}$	$\sqrt{2}$	1
60	$\dfrac{\pi}{3}$	$\dfrac{\sqrt{3}}{2}$	$\dfrac{1}{2}$	$\sqrt{3}$	$\dfrac{2}{\sqrt{3}}$	2	$\dfrac{1}{\sqrt{3}}$
90	$\dfrac{\pi}{2}$	1	0	undefined	1	undefined	0
120	$\dfrac{2\pi}{3}$	$\dfrac{\sqrt{3}}{2}$	$-\dfrac{1}{2}$	$-\sqrt{3}$	$\dfrac{2}{\sqrt{3}}$	-2	$-\dfrac{1}{\sqrt{3}}$
135	$\dfrac{3\pi}{4}$	$\dfrac{1}{\sqrt{2}}$	$-\dfrac{1}{\sqrt{2}}$	-1	$\sqrt{2}$	$-\sqrt{2}$	-1
150	$\dfrac{5\pi}{6}$	$\dfrac{1}{2}$	$-\dfrac{\sqrt{3}}{2}$	$-\dfrac{1}{\sqrt{3}}$	2	$-\dfrac{2}{\sqrt{3}}$	$-\sqrt{3}$
180	π	0	-1	0	undefined	-1	undefined
210	$\dfrac{7\pi}{6}$	$-\dfrac{1}{2}$	$-\dfrac{\sqrt{3}}{2}$	$\dfrac{1}{\sqrt{3}}$	-2	$-\dfrac{2}{\sqrt{3}}$	$\sqrt{3}$
225	$\dfrac{5\pi}{4}$	$-\dfrac{1}{\sqrt{2}}$	$-\dfrac{1}{\sqrt{2}}$	1	$-\sqrt{2}$	$-\sqrt{2}$	1
240	$\dfrac{4\pi}{3}$	$-\dfrac{\sqrt{3}}{2}$	$-\dfrac{1}{2}$	$\sqrt{3}$	$-\dfrac{2}{\sqrt{3}}$	-2	$\dfrac{1}{\sqrt{3}}$
270	$\dfrac{3\pi}{2}$	-1	0	undefined	-1	undefined	0
300	$\dfrac{5\pi}{3}$	$-\dfrac{\sqrt{3}}{2}$	$\dfrac{1}{2}$	$-\sqrt{3}$	$-\dfrac{2}{\sqrt{3}}$	2	$-\dfrac{1}{\sqrt{3}}$
315	$\dfrac{7\pi}{4}$	$-\dfrac{1}{\sqrt{2}}$	$\dfrac{1}{\sqrt{2}}$	-1	$-\sqrt{2}$	$\sqrt{2}$	-1
330	$\dfrac{11\pi}{6}$	$-\dfrac{1}{2}$	$\dfrac{\sqrt{3}}{2}$	$-\dfrac{1}{\sqrt{3}}$	-2	$\dfrac{2}{\sqrt{3}}$	$-\sqrt{3}$
360	2π	0	1	0	undefined	1	undefined

Two For You – Related Angles in Standard Position

1) Find the related angles in each of the other three quadrants for (a) $10°$ (b) $-25°$.

2) Find the related angles in each of the other three quadrants for (a) $\dfrac{\pi}{12}$ (b) $\dfrac{6\pi}{5}$.

Answers 1) (a) Quad 2: $170°$ Quad 3: $190°$ Quad 4: $-10°$
(b) Quad 1: $25°$ Quad 2: $155°$ Quad 3: $205°$

2)(a) Quad 2: $\dfrac{11\pi}{12}$ Quad 3: $\dfrac{13\pi}{12}$ Quad 4: $-\dfrac{\pi}{12}$

(b) Quad 1: $\dfrac{\pi}{5}$ Quad 2: $\dfrac{4\pi}{5}$ Quad 4: $-\dfrac{\pi}{5}$

Two For You – Trig Ratios
$(30°, 60°, 90°) \equiv (\pi/6, \pi/3, \pi/2)$ Triangle

1) State the cosecant, secant, and cotangent ratios for $30°$ and $60°$.
2) Is it a coincidence that $\sin(30°) = \cos(60°)$?

Answers

1) $\csc(30°) = \csc\left(\dfrac{\pi}{6}\right) = 2 \qquad \sec(30°) = \sec\left(\dfrac{\pi}{6}\right) = \dfrac{2}{\sqrt{3}} \qquad \cot(30°) = \cot\left(\dfrac{\pi}{6}\right) = \sqrt{3}$

$\csc(60°) = \csc\left(\dfrac{\pi}{3}\right) = \dfrac{2}{\sqrt{3}} \qquad \sec(60°) = \sec\left(\dfrac{\pi}{3}\right) = 2 \qquad \cot(60°) = \cot\left(\dfrac{\pi}{3}\right) = \dfrac{1}{\sqrt{3}}$

2) **NO! Opposite** for $30°$ is **adjacent** for $60°$! In fact, if a and b are two **complementary** angles in a triangle (that is, they add to $90°$), then $\sin a = \cos b$, $\cos a = \sin b$, and $\tan a = \cot b$.

Two For You – Trig Ratios
$(45°, 45°, 90°) \equiv (\pi/4, \pi/4, \pi/2)$ Triangle

1) State the cosecant, secant, and cotangent ratios for $45°$.

2) Using $\cos^2(\theta) = \dfrac{1 + \cos(2\theta)}{2}$, find the exact value of $\cos\left(\dfrac{\pi}{8}\right)$.

Answers 1) $\csc(45°) = \sec(45°) = \sqrt{2}$ $\cot(45°) = 1$ 2) $\cos\left(\dfrac{\pi}{8}\right) = \sqrt{\dfrac{2+\sqrt{2}}{4}}$

Two For You – Trig Ratios for $30°, 45°, 60°$ (and More)—A Table

In **absolute value**, the numbers (or ratios) in the 1) $30°$ row   2) $45°$ row
are the same as …

Answers 1) $150°, 210°, 330°$ rows   2) $135°, 225°, 315°$ rows

Trig Ratios for 30°, 45°, 60° (and More): A (Fabulous) Picture

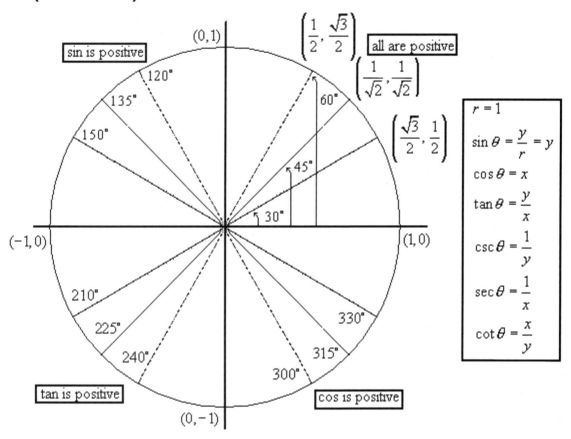

Example 1) In the above **fabulous** picture, look at the triangle formed by drawing a perpendicular from the 60° point to the x axis. Name the second, third, and fourth quadrant angles (between 90° and 360°) that give congruent triangles by drawing perpendiculars to the x axis.

Solution second quadrant: 120°; third quadrant: 240°; fourth quadrant: 300°

Example 2) Find, using the above **fabulous** picture, (a) $\sin(300°)$ (b) $\tan(180°)$.

Solution (a) 300° is a **fourth quadrant** angle, so the sine is **negative**. The corresponding first quadrant point is $\left(\frac{1}{2}, \frac{\sqrt{3}}{2}\right)$. Therefore, $\sin(300°) = -\frac{\sqrt{3}}{2}$.

(b) Using the point $(-1, 0)$, $\tan(180°) = \frac{y}{x} = \frac{0}{-1} = 0$.

Basic Trigonometric Graphs

$y = \sin x$

Domain $= \mathbb{R}$

Range $= [-1, 1]$

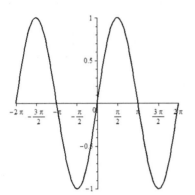

$y = \cos x$

Domain $= \mathbb{R}$

Range $= [-1, 1]$

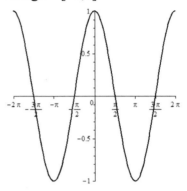

$y = \tan x$

Domain $= \left\{ x \in \mathbb{R} \mid x \neq \dfrac{\pi}{2} + k\pi,\ k \in \mathbb{Z} \right\}$

Range $= \mathbb{R}$

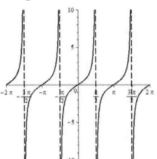

$y = \cot x$

Domain $= \left\{ x \in \mathbb{R} \mid x \neq k\pi,\ k \in \mathbb{Z} \right\}$

Range $= \mathbb{R}$

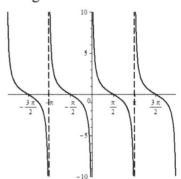

$y = \csc x$

Domain $= \left\{ x \in \mathbb{R} \mid x \neq k\pi,\ k \in \mathbb{Z} \right\}$

Range $= (-\infty, -1] \cup [1, \infty)$

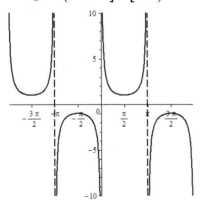

$y = \sec x$

Domain $= \left\{ x \in \mathbb{R} \mid x \neq \dfrac{\pi}{2} + k\pi,\ k \in \mathbb{Z} \right\}$

Range $= (-\infty, -1] \cup [1, \infty)$

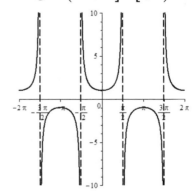

The Circle Definition of Sine and Cosine

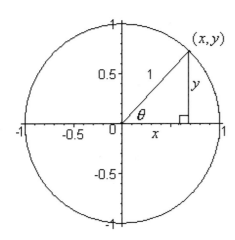

This is the circle $x^2 + y^2 = 1$. In the picture, θ is between 0 and $\pi/2$.

$\sin \theta = \dfrac{y}{1} = y$ and $\cos \theta = \dfrac{x}{1} = x$

Mathematicians, being sensible people, said, "Let θ be any angle. Draw this angle in standard postion (counter-clockwise from the positive x axis for positive angles, clockwise for negative). The terminal arm will puncture the circle at a point (x,y)."

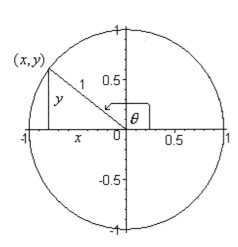

The mathematicians then said, "Let $\sin \theta = y$ and $\cos \theta = x$. The result: before, we had only sine and cosine for angles between 0 and $\pi/2$. Now, we have trig ratios for **all angles.** For any angle, the sine is the y value and the cosine is x. So, as θ goes from 0 to $\pi/2$ to π to $3\pi/2$ to 2π, the y value, that is, $\sin\theta$, goes from 0 to 1 to 0 to -1 to 0. The x value, that is, $\cos\theta$, goes from 1 to 0 to -1 to 0 to 1. From 2π to 4π, these patterns repeat. From 0 to -2π, they repeat in reverse. And that is all there really is to the sine and cosine functions."

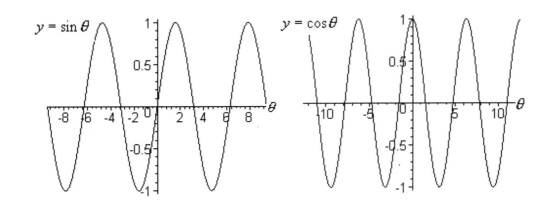

Solving the Trig Equation $\sin x = c$

DangerDangerDanger

If you ask your calculator (or math processor such as Maple) for help solving the equation $\sin x = c$ for x, **the calculator thinks this is the sine function!** →

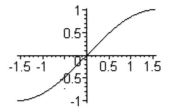

For each c between -1 and 1, the calculator reasons, "There is exactly **ONE** value of x between $-\pi/2$ and $\pi/2$ that makes $\sin x = c$. That value is the answer!" So your calculator gives you an answer in the **first quadrant when $c \geq 0$**, and the **fourth when $c < 0$**. But you and I know that $\sin x = c$ has an infinite number of solutions. Call the calculator answer the **principal** solution. Ask, when you solve $\sin x = c$, "Do I want the principal solution, another solution (eg., an answer in a different quadrant), or **all** possible solutions?"

Example 1) Solve, **using your knowledge of basic trig ratios (not a calculator!)**, giving, in radians, the principal solution, **a (not "the"!)** solution in the other **appropriate** quadrant, and the general solution: (a) $\sin x = 0.5$ (b) $\sin x = -0.5$

Solution (a) $\sin x = 0.5$ ∴ $x = \dfrac{\pi}{6}$ Using the CAST rule, sine is also positive in the second quadrant. A **(not "the")** corresponding second quadrant angle is $\pi - \dfrac{\pi}{6} = \dfrac{5\pi}{6}$. The general solution: $x = \dfrac{\pi}{6} + 2k\pi$ or $x = \dfrac{5\pi}{6} + 2k\pi$, for $k \in \mathbb{Z}$

(b) $\sin x = -0.5$ ∴ $x = -\dfrac{\pi}{6}$ Using the CAST rule, sine is also negative in the third quadrant. A **(not "the"!)** corresponding third quadrant angle is $\pi + \dfrac{\pi}{6} = \dfrac{7\pi}{6}$.

The general solution: $x = -\dfrac{\pi}{6} + 2k\pi$ or $x = \dfrac{7\pi}{6} + 2k\pi$, for $k \in \mathbb{Z}$

Example 2) Find the principal solution to (a) $\sin x = 0.1$ (b) $\sin x = -0.3$.

Solution (a) $\boxed{\text{MOST CALCULATORS} \\ 0.1 \text{ second function sin} =}$ $x \doteq 5.74°$ or 0.1 radians

(b) $\boxed{\text{MOST CALCULATORS} \\ 0.3 \pm \text{ second function sin} =}$ $x \doteq -17.5°$ or -0.305 radians

Two For You – Trig Ratios for 30°, 45°, 60° (and More)

Find: 1) $\cos(135°)$ 2) $\cot(-180°)$

Answers 1) $-\dfrac{1}{\sqrt{2}}$ 2) undefined

Two For You – Basic Trigonometric Graphs

1) Using "...", write the restrictions on the domain for the tangent and secant functions.

2) Using "...", write the restrictions on the domain for the cotangent and cosecant functions.

Answers 1) $...-\dfrac{3\pi}{2}, -\dfrac{\pi}{2}, \dfrac{\pi}{2}, \dfrac{3\pi}{2},...$ 2) $...-2\pi, -\pi, 0, \pi, 2\pi, ...$

One For You – The Circle Definition of Sine and Cosine

1) Use the circle definition of the trigonometric functions to explain the **CAST RULE**.

Answer 1) For example, in the second quadrant ($\pi/2 < \theta < \pi$) where $x < 0$ and $y > 0$, $\cos\theta = x$ is negative, $\sin\theta = y$ is positive, and $\tan\theta = \dfrac{y}{x}$ is negative. This explains the "S" in CA$\boxed{\text{S}}$T.

Two For You – Solving the Trig Equation $\sin x = c$

1) Find, in degree measure, the principal, the "other quadrant", and the general solution to $\sin x = \dfrac{1}{\sqrt{2}}$.

2) Find the principal and third quadrant solutions in radians to $\sin x = -0.9$.
   (Hint: for the third quadrant, use $\pi + |\text{principal solution}|$.)

Answers

1) principal solution: $x = 45°$; second quadrant: $x = 135°$;
general: $x = 45° + 360k°$ or $x = 135° + 360k°$, $k \in \mathbb{Z}$

2) principal solution: $x \doteq -1.1$; third quadrant: $x \doteq 4.3$

Solving the Trig Equation $\cos x = c$
DangerDangerDanger

If you ask your calculator for help in solving the equation $\cos x = c$ for x, **the calculator thinks this is the cosine function!** →

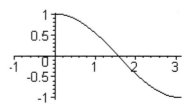

For each choice of c between -1 and 1, the calculator reasons, "There is **ONE** x between 0 and π that satisfies $\cos x = c$. That is the answer!" So your calculator gives you an answer in the **first quadrant when $c \geq 0$**, and the **second when $c < 0$**. But we know $\cos x = c$ **has an infinite number of solutions.** Call the calculator answer the **principal** solution. Ask, when you solve $\cos x = c$, "Do I want the principal solution, another solution (eg., an answer in a different quadrant), or **all** possible solutions?"

Example 1) Solve, **using your knowledge of basic trig ratios (not a calculator!)**, give the principal solution, **a (not "the"!) solution in the other appropriate quadrant**, and the general solution: (a) $\cos x = 0.5$ (b) $\cos x = -0.5$

Solution (a) $\cos x = 0.5$ ∴ $x = \dfrac{\pi}{3}$ Using the CAST rule, cosine is also positive in the fourth quadrant. A (**not "the"!**) corresponding fourth quadrant angle is $-\dfrac{\pi}{3}$.

The general solution: $x = \dfrac{\pi}{3} + 2k\pi$ or $x = -\dfrac{\pi}{3} + 2k\pi$, for $k \in \mathbb{Z}$

(b) $\cos x = -0.5$ ∴ $x = \dfrac{2\pi}{3}$ Using the CAST rule, cosine is also negative in the third quadrant. A (**not "the"!**) corresponding third quadrant angle is $2\pi - \dfrac{2\pi}{3} = \dfrac{4\pi}{3}$.

The general solution: $x = \dfrac{2\pi}{3} + 2k\pi$ or $x = \dfrac{4\pi}{3} + 2k\pi$, for $k \in \mathbb{Z}$

Example 2) Find the principal solution to (a) $\cos x = 0.1$ (b) $\cos x = -0.3$.

Solution (a) | MOST CALCULATORS |
| 0.1 second function cos = | $x \doteq 84.3°$ or 1.47 radians

(b) | MOST CALCULATORS |
| 0.3 ± second function cos = | $x \doteq 107.5°$ or 1.88 radians

The Sine Law

Check whether your calculator is in degree or radian measure!

According to the Sine Law, in the triangle at the right, $\dfrac{a}{\sin A} = \dfrac{b}{\sin B} = \dfrac{c}{\sin C}$.

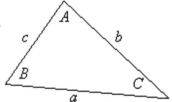

To use the Sine Law, you need two sides and a **non-contained angle** or any two angles and one side. (Remember: if you have two angles, you have three!) In the case of two sides and the contained angle, you can use the Cosine Law.

Example 1) If in the above triangle, $a = 12$, $b = 10$, and $A = \dfrac{\pi}{3}$, find B (in radians).

Solution $\dfrac{\sin B}{b} = \dfrac{\sin A}{a} \overset{\boxed{A=\frac{\pi}{3} \text{ and } a=12}}{=} \dfrac{\left(\frac{\sqrt{3}}{2}\right)}{12}$ and so $\sin B \overset{\boxed{b=10}}{=} \dfrac{10\sqrt{3}}{24} \doteq 0.722$

Therefore, $B \overset{\boxed{\text{MOST CALCULATORS} \atop \text{0.722 second funtion sin}}}{\doteq} 0.81$ radians.

Example 2) There are two possible triangles where $b = 11$, $c = 6$, and $C = 30°$, as illustrated at right. Find, in degrees, the **acute** and **obtuse** values of B.

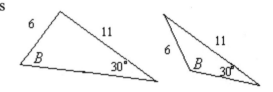

Solution $\dfrac{\sin B}{11} = \dfrac{\sin 30°}{6}$ and so $\sin B = \dfrac{(1/2)(11)}{6} = \dfrac{11}{12}$.

Therefore, $B \overset{\boxed{\text{If } B \text{ is acute!}}}{\doteq} 66.4°$ **OR** $B \overset{\boxed{\text{If } B \text{ is obtuse!}}}{\doteq} 180° - 66.4° = 113.6°$.

The Cosine Law

Check whether your calculator is in degree or radian measure!

According to the Cosine Law, in the triangle at the right,
$a^2 = b^2 + c^2 - 2bc \cos A$,
with similar formulas for b and c. Also,
$\cos A = \dfrac{b^2 + c^2 - a^2}{2bc}$, again with similar
formulas for $\cos B$ and $\cos C$.

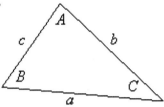

So, using the Cosine Law, to find a side you need the other two sides and the contained angle. (Compare the Sine Law). To find an angle, you need all three sides.

Example 1) In the above triangle, $a = 10$, $b = 12$, and $C = \dfrac{\pi}{7}$. Find c.

Solution $c^2 = 10^2 + 12^2 - 2(10)(12)\cos\left(\dfrac{\pi}{7}\right) \doteq 244 - 240(0.901) = 27.77$

and so $c \doteq 5.27$.

Example 2) Find, in degree measure, A in this triangle.

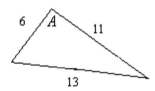

Solution $\cos A = \dfrac{11^2 + 6^2 - 13^2}{2 \cdot 11 \cdot 6} \doteq -0.0909$.

Therefore, A <u>MOST CALCULATORS: 0.0909 ± second function cos</u> $\doteq$ 95.2°.

Note: with the Cosine Law, if the required angle is obtuse (that is, between 90° and 180°), the calculator gives us the correct value. There is no possible ambiguity as can happen with the Sine Law. (See The Sine Law, Example 2, on page 136.)

Commonly Used Trigonometric Formulas Including Derivatives and Integrals

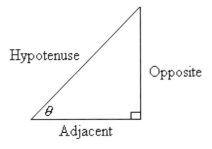

$$\sin \theta = \frac{\text{Opposite}}{\text{Hypotenuse}} = \frac{1}{\csc \theta} = \text{SOH}$$

$$\cos \theta = \frac{\text{Adjacent}}{\text{Hypotenuse}} = \frac{1}{\sec \theta} = \text{CAH}$$

$$\tan \theta = \frac{\text{Opposite}}{\text{Adjacent}} = \frac{1}{\cot \theta} = \text{TOA}$$

SOHCAHTOA

$\sin(-\theta) = -\sin \theta \qquad \cos(-\theta) = \cos \theta \qquad \tan(-\theta) = -\tan \theta$

$\cos^2 \theta + \sin^2 \theta = 1 \qquad 1 + \tan^2 \theta = \sec^2 \theta \qquad \cot^2 \theta + 1 = \csc^2 \theta$

$\sin(A \pm B) = \sin A \cos B \pm \cos A \sin B \qquad \cos(A \pm B) = \cos A \cos B \mp \sin A \sin B$

$\sin(2A) = 2 \sin A \cos A \qquad \cos(2A) = \cos^2 A - \sin^2 A$

$\tan(A \pm B) = \dfrac{\tan A \pm \tan B}{1 \mp \tan A \tan B} \qquad \tan(2A) = \dfrac{2 \tan A}{1 - \tan^2 A}$

$\sin^2 \theta = \dfrac{1 - \cos(2\theta)}{2} \qquad \cos^2 \theta = \dfrac{1 + \cos(2\theta)}{2}$

$\dfrac{d(\sin \theta)}{d\theta} = \cos \theta \qquad \int \sin \theta \, d\theta = -\cos \theta + C \quad$ (where C is a constant)

$\dfrac{d(\cos \theta)}{d\theta} = -\sin \theta \qquad \int \cos \theta \, d\theta = \sin \theta + C$

$\dfrac{d(\tan \theta)}{d\theta} = \sec^2 \theta \qquad \int \tan \theta \, d\theta = -\ln |\cos \theta| + C$

$\dfrac{d(\csc \theta)}{d\theta} = -\csc \theta \cot \theta \qquad \int \csc \theta \, d\theta = \ln |\csc \theta - \cot \theta| + C$

$\dfrac{d(\sec \theta)}{d\theta} = \sec \theta \tan \theta \qquad \int \sec \theta \, d\theta = \ln |\sec \theta + \tan \theta| + C$

$\dfrac{d(\cot \theta)}{d\theta} = -\csc^2 \theta \qquad \int \cot \theta \, d\theta = \ln |\sin \theta| + C$

Two For You – Solving the Trig Equation $\cos x = c$

1) Find, in degrees, the principal, the "other quadrant", and the general solution to $\cos x = \dfrac{1}{\sqrt{2}}$.

2) Find the principal and third quadrant solutions in radians to $\cos x = -0.9$.
(Hint: for the third quadrant, use $2\pi - |\text{principal solution}|$.)

Answers

1) principal solution: $x = 45°$; fourth quadrant: $x = -45°$;

general: $x = 45° + 360k$ or $x = -45° + 360k$, $k \in \mathbb{Z}$

2) principal solution: $x \doteq 2.69$; third quadrant: $x \doteq 3.59$

One For You – The Sine Law

1) In the triangle below, $a = 3$, $c = 5$, and $A = 30°$.
Find in degrees the two possible values of C.

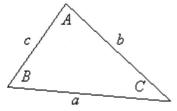

Answer 1) $C \doteq 56.44°$ [If C is acute!] **OR** $C \doteq 123.56°$ [If C is obtuse!]

One For You – The Cosine Law

1) In this triangle, use the Cosine Law to find side b and then, in degrees, A.

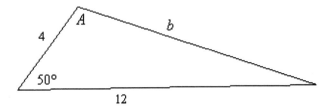

Answer 1) $b \doteq 9.9$ and $A \doteq 112°$

Two For You – Commonly Used Trigonometric Formulas

How can you derive the formulas 1) $1 + \tan^2 \theta = \sec^2 \theta$ and 2) $\cot^2 \theta + 1 = \csc^2 \theta$ from $\cos^2 \theta + \sin^2 \theta = 1$?

Answers 1) Divide the formula $\cos^2 \theta + \sin^2 \theta = 1$ by $\cos^2 \theta$.
2) Divide the formula $\cos^2 \theta + \sin^2 \theta = 1$ by $\sin^2 \theta$.

Basic Inverse Trigonometric Graphs

With trigonometric functions, we input angles and the outputs are ratios. With inverse trigonometric functions this is reversed: we input ratios and the outputs are angles.

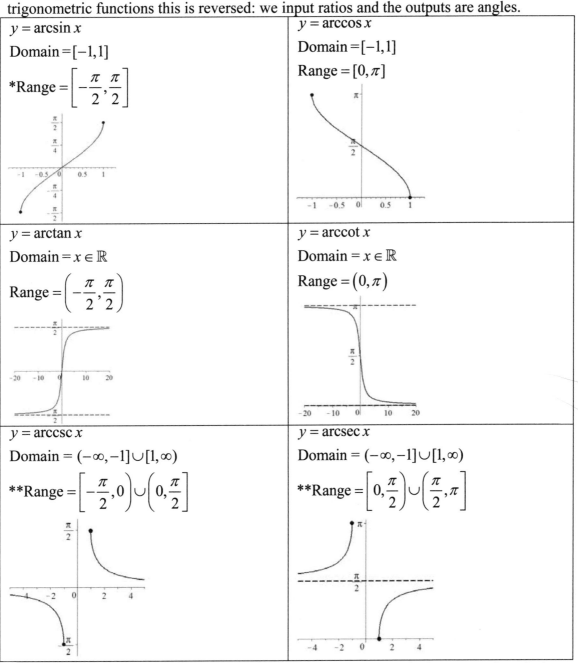

$y = \arcsin x$	$y = \arccos x$
Domain $= [-1, 1]$	Domain $= [-1, 1]$
*Range $= \left[-\dfrac{\pi}{2}, \dfrac{\pi}{2} \right]$	Range $= [0, \pi]$

$y = \arctan x$	$y = \text{arccot}\, x$
Domain $= x \in \mathbb{R}$	Domain $= x \in \mathbb{R}$
Range $= \left(-\dfrac{\pi}{2}, \dfrac{\pi}{2} \right)$	Range $= (0, \pi)$

$y = \text{arccsc}\, x$	$y = \text{arcsec}\, x$
Domain $= (-\infty, -1] \cup [1, \infty)$	Domain $= (-\infty, -1] \cup [1, \infty)$
**Range $= \left[-\dfrac{\pi}{2}, 0 \right) \cup \left(0, \dfrac{\pi}{2} \right]$	**Range $= \left[0, \dfrac{\pi}{2} \right) \cup \left(\dfrac{\pi}{2}, \pi \right]$

* Note 1: The ranges are restricted to two quadrants. For example the range of arcsin is $\left[-\dfrac{\pi}{2}, \dfrac{\pi}{2} \right]$ to ensure there is only one y value for each x.

** Note 2: some authors choose to use the first and third quadrant angles for the range of the arccsc and arcsec functions.

Easy Limits: "No Problem" Problems

When evaluating $\lim_{x \to a} f(x)$, **YOU MUST NOT LET** x **EQUAL** a. You consider instead the "behaviour" of the function $f(x)$ as x gets closer and closer to a. However, we all know that as long as the function is "well-behaved", we do, in practice, just substitute a in for x in $f(x)$.

Example 1) Evaluate the following limits:

(a) $\lim_{x \to 3} x^2$ (b) $\lim_{x \to \frac{\pi}{4}} \sin x$ (c) $\lim_{x \to 3} \left(\dfrac{x^2 - 1}{\ln x} \right)$

Solution

(a) $\lim_{x \to 3} x^2 = 9$ (b) $\lim_{x \to \frac{\pi}{4}} \sin x = \sin\left(\dfrac{\pi}{4}\right) = \dfrac{1}{\sqrt{2}}$ (c) $\lim_{x \to 3} \left(\dfrac{x^2 - 1}{\ln x} \right) = \dfrac{8}{\ln 3}$

Example 2) Let $g(x) = \begin{cases} x^2 - 1, & \text{if } x < -4 \\ x^3, & \text{if } x \geq -4 \end{cases}$.

Evaluate: (a) $\lim_{x \to -6} g(x)$ (b) $\lim_{x \to -2} g(x)$ (c) $\lim_{x \to -4.01} g(x)$ (d) $\lim_{x \to -3.99} g(x)$

Solution (a) $\lim_{x \to -6} g(x)$ $\boxed{x < -4}$ $= 35$ (b) $\lim_{x \to -2} g(x)$ $\boxed{x > -4}$ $= -8$

(c) $\lim_{x \to -4.01} g(x)$ $\boxed{x \text{ is still LESS than } -4.}$ $= (-4.01)^2 - 1 = 15.0801$

(d) $\lim_{x \to -3.99} g(x)$ $\boxed{x \text{ is still GREATER than } -4.}$ $= (-3.99)^3 = 63.521199$

> We have to be careful with $g(x)$ only with the limit as x approaches **EXACTLY** -4.

Example 3) Evaluate: (a) $\lim_{x \to 2.3} [[x]]$ (b) $\lim_{x \to -3.1} [[x]]$

Solution (a) $\lim_{x \to 2.3} [[x]] = [[2.3]] = 2$ (b) $\lim_{x \to -3.1} [[x]] = [[-3.1]] = -4$

We must be careful with the greatest integer function $[[x]]$ only when x is approaching an integer.

"0/0" Limits

We can't divide by 0! However, if we try to evaluate $\lim_{x \to a}\left(\dfrac{f(x)}{g(x)}\right)$ by simply substituting $x = a$ (treating the limit as a **"No Problem" Problem**—see page 142) and the result is "0/0", then, often, we can find a factor of $x - a$ in both the numerator and the denominator. It is this factor that is causing both the top and the bottom to approach 0 as $x \to a$. So factor the top, factor the bottom. If you do find a factor on both the top and the bottom of $x - a$, divide it out. Remember, $x \neq a$ and so you are **not** dividing by 0. Now, fingers crossed, you might just have a **"No Problem" Problem**!

Example 1) Evaluate the following limits:

(a) $\lim\limits_{x \to 3}\left(\dfrac{x^2 - 9}{x - 3}\right)$ (b) $\lim\limits_{x \to -1}\left(\dfrac{x^2 + 2x + 1}{x^3 + 1}\right)$ (c) $\lim\limits_{x \to 16}\left(\dfrac{\sqrt{x} - 4}{x - 16}\right)$ (d) $\lim\limits_{x \to -125}\left(\dfrac{x + 125}{x^{1/3} + 5}\right)$

Solution (a) $\lim\limits_{x \to 3}\left(\dfrac{x^2 - 9}{x - 3}\right)$ [Factor the top and then divide out the common $(x-3)$. Say **GOODBYE** to the problem "0/0"!] $= \lim\limits_{x \to 3}\left(\dfrac{(x-3)(x+3)}{x-3}\right) = \lim\limits_{x \to 3}(x + 3) = 6$

(b) $\lim\limits_{x \to -1}\left(\dfrac{x^2 + 2x + 1}{x^3 + 1}\right) = \lim\limits_{x \to -1}\left(\dfrac{(x+1)^2}{(x+1)(x^2 - x + 1)}\right) = \lim\limits_{x \to -1}\left(\dfrac{x+1}{x^2 - x + 1}\right) = \dfrac{0}{3} = 0$

(c) $\lim\limits_{x \to 16}\left(\dfrac{\sqrt{x} - 4}{x - 16}\right)$ [Rationalize the top using difference of squares: $(a-b)(a+b) = a^2 - b^2$, where $a = \sqrt{x}$ and $b = 4$.] $= \lim\limits_{x \to 16}\left(\dfrac{\sqrt{x} - 4}{x - 16}\right)\left(\dfrac{\sqrt{x} + 4}{\sqrt{x} + 4}\right)$

$= \lim\limits_{x \to 16}\left(\dfrac{x - 16}{(x - 16)(\sqrt{x} + 4)}\right) = \lim\limits_{x \to 16}\left(\dfrac{1}{\sqrt{x} + 4}\right) = \dfrac{1}{8}$

OR

(c) $\lim\limits_{x \to 16}\left(\dfrac{\sqrt{x} - 4}{x - 16}\right)$ [Factor the bottom using difference of squares.] $= \lim\limits_{x \to 16}\left(\dfrac{\sqrt{x} - 4}{(\sqrt{x} - 4)(\sqrt{x} + 4)}\right) = \lim\limits_{x \to 16}\left(\dfrac{1}{\sqrt{x} + 4}\right) = \dfrac{1}{8}$

(d) $\lim\limits_{x \to -125}\left(\dfrac{x + 125}{x^{1/3} + 5}\right)$ [Factor the top using sum of cubes: $a^3 + b^3$ where $a = x^{1/3}$ and $b = 5$.] $= \lim\limits_{x \to -125}\left(\dfrac{(x^{1/3} + 5)(x^{2/3} - 5x^{1/3} + 25)}{x^{1/3} + 5}\right)$

$= \lim\limits_{x \to -125}\left(x^{2/3} - 5x^{1/3} + 25\right) = 75$

One-Sided Limits

If a function changes its definition at a point, we must check the limits from the left and the right separately.

Example 1) Let $f(x) = \begin{cases} 3, & \text{if } x < 1 \\ 0, & \text{if } x = 1 \\ x+2, & \text{if } x > 1 \end{cases}$ and $g(x) = \begin{cases} \cos(\pi x), & \text{if } x < 1 \\ 5, & \text{if } x = 1 \\ \sin(\pi x), & \text{if } x > 1 \end{cases}$.

Evaluate: (a) $\lim\limits_{x \to 1^-} f(x)$ (b) $\lim\limits_{x \to 1^+} f(x)$ (c) $\lim\limits_{x \to 1} f(x)$ (d) $\lim\limits_{x \to 1^-} g(x)$

(e) $\lim\limits_{x \to 1^+} g(x)$ (f) $\lim\limits_{x \to 1} g(x)$

Solution (a) $\lim\limits_{x \to 1^-} f(x) \overset{\boxed{x<1}}{=} \lim\limits_{x \to 1^-}(3) = 3$

(b) $\lim\limits_{x \to 1^+} f(x) \overset{\boxed{x>1}}{=} \lim\limits_{x \to 1^+}(x+2) = 3$

(c) $\lim\limits_{x \to 1} f(x) = 3$ since the left-hand limit and the right-hand limit are both equal to 3.

(d) $\lim\limits_{x \to 1^-} g(x) \overset{\boxed{x<1}}{=} \lim\limits_{x \to 1^-} \cos(\pi x) = \cos(\pi) = -1$

(e) $\lim\limits_{x \to 1^+} g(x) \overset{\boxed{x>1}}{=} \lim\limits_{x \to 1^+} \sin(\pi x) = \sin(\pi) = 0$

(f) $\lim\limits_{x \to 1} g(x)$ does not exist since the left-hand limit ≠ the right-hand limit.

Note that $f(1) = 0$ and $g(1) = 5$ and **neither plays ANY part in the limits!**
Sometimes, YOU must realize that one-sided limits are necessary!

Example 2) Evaluate the following limits:

(a) $\lim\limits_{x \to 2} \dfrac{|x-2|}{x-2}$ (b) $\lim\limits_{x \to 4}(x + [[x]])$

Solution (a) $\lim\limits_{x \to 2^+} \dfrac{|x-2|}{x-2} \overset{\boxed{\substack{x>2 \text{ so } x-2>0 \\ \therefore |x-2|=x-2}}}{=} \lim\limits_{x \to 2^+} \dfrac{x-2}{x-2} = \lim\limits_{x \to 2^+} 1 = 1$

$\lim\limits_{x \to 2^-} \dfrac{|x-2|}{x-2} \overset{\boxed{\substack{x<2 \text{ so } x-2<0 \\ \therefore |x-2|=-(x-2)}}}{=} \lim\limits_{x \to 2^-} \dfrac{-(x-2)}{x-2} = \lim\limits_{x \to 2^-}(-1) = -1 \quad \therefore \lim\limits_{x \to 2} \dfrac{|x-2|}{x-2}$ does not exist.

(b) $\lim\limits_{x \to 4^+}(x + [[x]]) \overset{\boxed{\substack{4<x<5 \text{ so} \\ [[x]]=4}}}{=} \lim\limits_{x \to 4^+}(x+4) = 8 \qquad \lim\limits_{x \to 4^-}(x + [[x]]) \overset{\boxed{\substack{3<x<4 \text{ so} \\ [[x]]=3}}}{=} \lim\limits_{x \to 4^-}(x+3) = 7$

$\therefore \lim\limits_{x \to 4}(x + [[x]])$ does not exist.

Two For You – Basic Inverse Trigonometric Graphs

Note that $\arcsin\left(\sin\left(\frac{\pi}{2}\right)\right) = \arcsin(1) = \frac{\pi}{2}$. However $\arcsin\left(\sin\left(\frac{5\pi}{2}\right)\right) = \arcsin(1) = \frac{\pi}{2}$.

Thinking of domain and ranges, when does
1) a) $\arcsin(\sin(\theta)) = \theta$? b) $\sin(\arcsin(x)) = x$?
2) a) $\text{arctrig}(\text{trig}(\theta)) = \theta$? b) $\text{trig}(\text{arctrig}(x)) = x$?

Answers 1)(a) $\forall \, \theta \in$ Range of arcsin, that is, $\theta \in \left[-\frac{\pi}{2}, \frac{\pi}{2}\right]$ *

(b) $\forall \, x \in$ Domain of arcsin, that is, $x \in [-1, 1]$.

$\arcsin(x)$ will be an angle in $[-\pi/2, \pi/2]$, sine will take in this angle and give back the original ratio x.

2)(a) $\forall \, \theta \in$ Domain of trig **and** $\theta \in$ Range of arctrig

(b) $\forall \, x \in$ Domain of arctrig

* $\forall$ means "for all"

Two For You – Easy Limits: "No Problem" Problems

1) Evaluate the following limits:

(a) $\lim\limits_{x \to \pi} \cos(2x)$ (b) $\lim\limits_{x \to -4.2} [[x]]$ (c) $\lim\limits_{x \to 3} \frac{|x-5|}{x-5}$

2) Let $f(x) = \begin{cases} e^x, & \text{if } x \neq 2 \\ 5, & \text{if } x = 2 \end{cases}$. Find $\lim\limits_{x \to 2} f(x)$.

Answers 1)(a) 1   (b) -5 (c) -1 2) e^2

Three For You – "0/0" Limits

Evaluate the following limits:

1) $\lim\limits_{x \to 10} \left(\dfrac{x^2 - 100}{x - 10} \right)$
2) $\lim\limits_{x \to -2} \left(\dfrac{x^3 + 8}{x^2 - 4} \right)$
3) $\lim\limits_{x \to -27} \left(\dfrac{x^{1/3} + 3}{x + 27} \right)$

Answers 1)(a) 20   2) -3 3) $\dfrac{1}{27}$

Two For You – One-Sided Limits

1) Let $f(x) = \begin{cases} x^2, & \text{if } x < -4 \\ x^3, & \text{if } x \geq -4 \end{cases}$. Find $\lim\limits_{x \to -4^-} f(x)$.

2) Evaluate: (a) $\lim\limits_{x \to -3^+} \dfrac{x+3}{|x+3|}$ (b) $\lim\limits_{x \to -5^-} (x[[x]])$

Answers 1) 16   2)(a) 1   (b) 30

Limits which Approach $\pm\infty$

Whenever a limit yields the form $\dfrac{\text{"constant"}}{0}$ where the constant **IS NOT 0**, the answer **WILL BE INFINITE**. We just need to determine if it is $+\infty$ or $-\infty$.

Keep these basic infinite limits in mind: $\boxed{\lim\limits_{x\to 0^+}\dfrac{1}{x}=+\infty \text{ and } \lim\limits_{x\to 0^-}\dfrac{1}{x}=-\infty}$

Also, remember that some functions have infinite limits "built in".
For example, $\lim\limits_{x\to 0^+}\ln x=-\infty$.

Example 1) Evaluate (a) $\lim\limits_{x\to 3^+}\left(\dfrac{x^2+1}{x-3}\right)$ (b) $\lim\limits_{x\to 3^-}\left(\dfrac{x^2+1}{x-3}\right)$

Solution (a) Note that just substituting yields $\dfrac{\text{"}10\text{"}}{0}$. Since x is **greater than** and approaching 3, therefore $x-3$ is **POSITIVE** and **SMALL**.

$\therefore \lim\limits_{x\to 3^+}\left(\dfrac{x^2+1}{x-3}\right) = 10\lim\limits_{x\to 3^+}\left(\dfrac{1}{x-3}\right) = +\infty$

(See the personal note below to explain where the "10" comes from!)

> **A personal note :** in these questions, I like to evaluate everything except the factor causing the "0 denominator" and pull the result OUTSIDE the limit.

(b) $\lim\limits_{x\to 3^-}\left(\dfrac{x^2+1}{x-3}\right) = 10\lim\limits_{x\to 3^-}\left(\dfrac{1}{x-3}\right) = -\infty$

Example 2) Evaluate $\lim\limits_{x\to 4^-}\left(\dfrac{1-x^2}{(4-x)(x+1)}\right)$.

Solution $\lim\limits_{x\to 4^-}\left(\dfrac{1-x^2}{(4-x)(x+1)}\right) = \dfrac{-15}{5}\lim\limits_{x\to 4^-}\left(\dfrac{1}{-(x-4)}\right)$

(Evaluate everything but the $4-x$ term and rewrite it as $-(x-4)$.)

$= 3\lim\limits_{x\to 4^-}\left(\dfrac{1}{x-4}\right) = -\infty$

(Why? It's easier to look at $x-4$ than $4-x$ as $x\to 4^-$.)

($\because x<4 \therefore x-4<0$)

Limits at Infinity

Keep the basic limits "at infinity" (ie., limits where x approaches $+\infty$ or $-\infty$) in mind:

$$\lim_{x \to +\infty} \frac{1}{x} = 0 \quad \text{and} \quad \lim_{x \to -\infty} \frac{1}{x} = 0$$

Some functions have limits at infinity "built in". For example, $\lim_{x \to -\infty} 2^x = 0$.

Often, these kinds of questions arise when you have $\dfrac{\text{(almost) a polynomial}}{\text{(almost) another polynomial}}$.

In these examples, the **EASIEST** method is to divide the top and the bottom by the **HIGHEST POWER OF x IN THE DENOMINATOR!**

Example 1) Evaluate:

(a) $\lim_{x \to \infty} \left(\dfrac{x+1}{3x+2} \right)$ (b) $\lim_{x \to -\infty} \left(\dfrac{x^2 + \sin x}{x^3 + 2x} \right)$ (c) $\lim_{x \to -\infty} \left(\dfrac{x^3 - x + \cos x}{x^2 + 1 + \sin x} \right)$

Solution (a) $\lim_{x \to \infty} \left(\dfrac{x+1}{3x+2} \right) \overset{\text{Divide top and bottom by } x.}{=} \lim_{x \to \infty} \dfrac{\left(\dfrac{x+1}{x}\right)}{\left(\dfrac{3x+2}{x}\right)} = \lim_{x \to \infty} \left(\dfrac{1 + \dfrac{1}{x}}{3 + \dfrac{2}{x}} \right) = \dfrac{1+0}{3+0} = \dfrac{1}{3}$

(b) $\lim_{x \to -\infty} \left(\dfrac{x^2 + \sin x}{x^3 + 2x} \right) \overset{\text{Divide top and bottom by } x^3.}{=} \lim_{x \to -\infty} \left(\dfrac{\dfrac{1}{x} + \dfrac{\sin x}{x^3}}{1 + \dfrac{2}{x^2}} \right)$

> Remember $-1 \leq \sin x \leq 1$, so $\dfrac{\sin x}{x^3}$ is, in magnitude, $\dfrac{\text{small number}}{\text{BIG NUMBER}}$ and will therefore approach 0!

$= \dfrac{0+0}{1+0} = \dfrac{0}{1} = 0$

(c) $\lim_{x \to -\infty} \left(\dfrac{x^3 - x + \cos x}{x^2 + 1 + \sin x} \right) \overset{\text{Divide top and bottom by } x^2, \text{ NOT } x^3.}{=} \lim_{x \to -\infty} \left(\dfrac{x - \dfrac{1}{x} + \dfrac{\cos x}{x^2}}{1 + \dfrac{1}{x^2} + \dfrac{\sin x}{x^2}} \right)$

> Watch the special way we handle the x on top. Evaluate each limit except the one that still goes to $\pm\infty$.

$= \dfrac{\lim_{x \to -\infty} x - 0 + 0}{1 + 0 + 0} = -\infty$

An "∞ − ∞" Limit: $\lim\limits_{x \to \infty} \left(\sqrt{x^2+8x}-x\right)$

You might think "$\infty - \infty$" has to be 0. Not necessarily. In fact, anything can happen, depending on how the first term approaches ∞ compared to the second term.

For example, $\lim\limits_{x \to \infty}(x^2 - x) = \infty \qquad \lim\limits_{x \to \infty}(x - x) = 0 \qquad \lim\limits_{x \to \infty}(x - x^2) = -\infty$

Example 1) Evaluate $\lim\limits_{x \to \infty}\left(\sqrt{x^2+8x}-x\right)$.

Solution $\lim\limits_{x \to \infty}\left(\sqrt{x^2+8x}-x\right)$

Rationalize the numerator using $(a-b)(a+b)=a^2-b^2$, with $a=\sqrt{x^2+8x}$ and $b=x$.

$= \lim\limits_{x \to \infty}\left(\sqrt{x^2+8x}-x\right)\left(\dfrac{\sqrt{x^2+8x}+x}{\sqrt{x^2+8x}+x}\right)$

$= \lim\limits_{x \to \infty}\left(\dfrac{x^2+8x-x^2}{\sqrt{x^2+8x}+x}\right) = \lim\limits_{x \to \infty}\left(\dfrac{8x}{\sqrt{x^2+8x}+x}\right)$

Divide the top and the bottom by x. Use $x=\sqrt{x^2}$ (which is true when $x>0$) to divide x into $\sqrt{x^2+8x}$.*

$= \lim\limits_{x \to \infty}\left(\dfrac{8}{\sqrt{1+\dfrac{8}{x}}+1}\right) = \dfrac{8}{2} = 4$

BUT...

Example 2) Evaluate $\lim\limits_{x \to \infty}\left(\sqrt{x^2+8}-x\right)$

Solution $\lim\limits_{x \to \infty}\left(\sqrt{x^2+8}-x\right)$

Rationalize the numerator using $(a-b)(a+b)=a^2-b^2$.

$= \lim\limits_{x \to \infty}\left(\sqrt{x^2+8}-x\right)\left(\dfrac{\sqrt{x^2+8}+x}{\sqrt{x^2+8}+x}\right)$

$= \lim\limits_{x \to \infty}\left(\dfrac{x^2+8-x^2}{\sqrt{x^2+8}+x}\right) = \lim\limits_{x \to \infty}\left(\dfrac{8}{\sqrt{x^2+8}+x}\right)$

Divide the top and the bottom by x. Use $x=\sqrt{x^2}$ (which is true when $x>0$.)

$= \lim\limits_{x \to \infty}\left(\dfrac{\dfrac{8}{x}}{\sqrt{1+\dfrac{8}{x}}+1}\right) = \dfrac{0}{2} = 0$

*$|x| = \sqrt{x^2} = \begin{cases} -x, & x < 0 \\ x, & x \geq 0 \end{cases}$

When $x \to \infty$, we can certainly say $x > 0$ and so $\sqrt{x^2} = x$.

Limits: A Summary

This is a non-standard page but students find this summary of limit questions **very** useful. Almost every limit question you encounter will look like one of these!

Case 1) "No Problem" Problems. Letting x get close to a leads to a clear answer.

Example 1) $\lim\limits_{x \to 6}\left(\dfrac{x^2 - 25}{x + 5}\right) = \dfrac{11}{11} = 1$

Case 2) "$\dfrac{c}{0}$", where $c \neq 0$. Here, the answer will be $+\infty$ or $-\infty$, or possibly both, in which case you need to consider one-sided limits.

Example 2) $\lim\limits_{x \to -5^+}\left(\dfrac{x^2}{x + 5}\right) = 25 \lim\limits_{x \to -5^+}\left(\dfrac{1}{x + 5}\right) \overset{\boxed{x+5>0}}{=} +\infty$

Case 3) "$\dfrac{0}{0}$" when $x = a$. Divide out $\dfrac{x - a}{x - a}$, possibly more than once, and the problem should reduce to case 1 or 2.

Example 3) $\lim\limits_{x \to -5}\left(\dfrac{x^2 - 25}{x + 5}\right) = \lim\limits_{x \to -5}\left(\dfrac{(x-5)\cancel{(x+5)}}{\cancel{x+5}}\right) = -10$

Case 4) One-sided limits. Any of cases 1, 2, and 3 can come in the form of one-sided limits. These can be either explicit or require you to consider separate cases.

Example 4) Explicit: $\lim\limits_{x \to 2^-}[[x]] = 1$;

Implicit: $\lim\limits_{x \to 2}[[x]] \because \lim\limits_{x \to 2^-}[[x]] = 1 \neq \lim\limits_{x \to 2^+}[[x]] = 2$, $\therefore \lim\limits_{x \to 2}[[x]]$ does not exist.

Case 5) x approaches $+\infty$ or $-\infty$ in a $\dfrac{\text{polynomial}}{\text{polynomial}}$. Divide the top and bottom by the highest power of x in the bottom, to obtain one of these outcomes:

(i) $\dfrac{0}{c}$ (ii) $\dfrac{b}{c}$ (iii) "$\pm\dfrac{\infty}{c}$", where $b, c \neq 0$.

Then the answer is (i) 0 (ii) $\dfrac{b}{c}$ (iii) $+\infty$ or $-\infty$

Example 5) $\lim\limits_{x \to \infty}\left(\dfrac{x^5 + 1}{x^4 + x^3}\right) \overset{\boxed{\text{Divide top and bottom by } x^4.}}{=} \lim\limits_{x \to \infty}\left(\dfrac{x + \dfrac{1}{x^4}}{1 + \dfrac{1}{x}}\right) = \dfrac{\lim\limits_{x \to \infty} x + 0}{1 + 0} = +\infty$

Two For You – Limits Which Approach $\pm\infty$

Find the following limits: 1)(a) $\lim\limits_{x \to 1^+} \ln(x-1)$ (b) $\lim\limits_{x \to \frac{\pi}{2}^-} \tan x$

2) $\lim\limits_{x \to 5^-} \left(\dfrac{x^2 - 25}{(x-5)^2} \right)$ (Hint: factor and simplify first.)

Answers 1)(a) $-\infty$ (b) $+\infty$ 2) $-\infty$

Two For You – Limits At Infinity

Evaluate the limits:

1) $\lim\limits_{x \to -\infty} \left(\dfrac{4x^3 - 2x + e^x}{3x^3 - 5x^2 + \sin x} \right)$

2) $\lim\limits_{x \to \infty} \left(\dfrac{x+1}{\sqrt{4x^2 + x} + 5x} \right)$

$\left(\begin{array}{l} \text{Hint: the highest power of } x \text{ in the bottom is } x, \text{ because of the "}\sqrt{}\text{".} \\ \text{Another hint: } \dfrac{\sqrt{4x^2+x}}{x} \overset{\boxed{\because x \to \infty\ \therefore\ x > 0 \text{ and so } x = \sqrt{x^2}}}{=} \dfrac{\sqrt{4x^2+x}}{\sqrt{x^2}} = \sqrt{\text{you get the idea!}} \end{array} \right)$

Answers 1) $\dfrac{4}{3}$ 2) $\dfrac{1}{7}$

Two For You – An "∞ − ∞" Limit: $\lim\limits_{x \to \infty}\left(\sqrt{x^2+8x}-x\right)$

1) Evaluate: $\lim\limits_{x \to \infty}\left(\sqrt{x^2-x+1}-x\right)$

2) Evaluate: $\lim\limits_{x \to -\infty}\left(\sqrt{x^2-3x}+x\right)$

$\left(\begin{array}{l}\text{Hint: For } x<0,\ \sqrt{x^2}=-x.\ \text{For example, if } x=-10,\\ \text{then } \sqrt{(-10)^2}=\sqrt{100}=10=-(-10).\\ \\ \text{Another hint: } \dfrac{\sqrt{x^2-3x}}{x} \overset{\boxed{\text{For } x<0,\ x=-\sqrt{x^2}}}{=} \dfrac{\sqrt{x^2-3x}}{-\sqrt{x^2}} = -\sqrt{\dfrac{x^2-3x}{x^2}} = -\sqrt{1-\dfrac{3}{x}}\end{array}\right)$

Answers 1) $-\dfrac{1}{2}$ 2) $\dfrac{3}{2}$

Two For You – Limits: A Summary

Evaluate: 1) $\lim\limits_{x \to 3^-}\left(\dfrac{3-4x}{x-3}\right)$ 2) $\lim\limits_{x \to -\infty}\left(\dfrac{x^5+1}{x^4+x^3}\right)$

Answers 1) ∞ 2) $-\infty$

Variations on $\lim_{\theta \to 0}\left(\dfrac{\sin\theta}{\theta}\right) = 1$

Lots of trigonometric "$\dfrac{0}{0}$" limits are evaluated with a little clever use of $\lim_{\theta \to 0} \dfrac{\sin\theta}{\theta} = 1$.

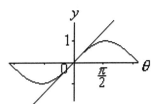

(The interpretation of this limit is very intuitive: when θ is small, $\sin\theta$ and θ are nearly equal. Therefore, this ratio approaches 1.)

Example 1) Evaluate the following limits:

(a) $\lim_{\theta \to 0} \dfrac{\theta}{\sin\theta}$ (b) $\lim_{\theta \to 0} \dfrac{\sin 3\theta}{\theta}$ (c) $\lim_{\theta \to 0} \dfrac{\tan 2\theta}{\sin\theta}$ (d) $\lim_{\theta \to 0} \dfrac{1-\cos\theta}{\theta}$

Solution (a) $\lim_{\theta \to 0} \dfrac{\theta}{\sin\theta} = \lim_{\theta \to 0} \dfrac{1}{\left(\dfrac{\sin\theta}{\theta}\right)} = \dfrac{1}{1} = 1$

(b) $\lim_{\theta \to 0} \dfrac{\sin 3\theta}{\theta}$ ["3θ" is playing the role of "θ" here! Put 3 in the bottom and compensate with 3 on top to keep things equal!] $= \lim_{\theta \to 0} \dfrac{3\sin 3\theta}{3\theta}$ [3θ approaches 0 as θ approaches 0.] $= \lim_{3\theta \to 0} \dfrac{3\sin 3\theta}{3\theta} = 3(1) = 3$

(c) $\lim_{\theta \to 0} \dfrac{\tan 2\theta}{\sin\theta}$ [Use $\tan 2\theta = \dfrac{\sin 2\theta}{\cos 2\theta}$.] $= \lim_{\theta \to 0} \dfrac{\sin 2\theta}{\sin\theta \cos 2\theta}$

[Throw in some "θ's" to make the variables MATCH!] $= \lim_{\theta \to 0} \left(\dfrac{\sin 2\theta}{2\theta}\right)\left(\dfrac{\theta}{\sin\theta}\right)\left(\dfrac{2}{\cos 2\theta}\right) = (1)(1)(2) = 2$

(d) $\lim_{\theta \to 0} \dfrac{1-\cos\theta}{\theta}$ [Use $\sin^2\theta = 1-\cos^2\theta$.] $= \lim_{\theta \to 0}\left(\dfrac{1-\cos\theta}{\theta}\right)\left(\dfrac{1+\cos\theta}{1+\cos\theta}\right) = \lim_{\theta \to 0}\left(\dfrac{1-\cos^2\theta}{\theta(1+\cos\theta)}\right)$

$= \lim_{\theta \to 0}\left(\dfrac{\sin^2\theta}{\theta(1+\cos\theta)}\right) = \lim_{\theta \to 0}\left(\dfrac{\sin\theta}{\theta}\right)\left(\dfrac{\sin\theta}{1+\cos\theta}\right) = (1)\left(\dfrac{0}{2}\right) = 0$

L'Hôpital's Rule

L'Hôpital's rule gives us a method for dealing with "0/0" or "∞/∞" limits. Let's say $\lim_{x \to a} f(x) = \lim_{x \to a} g(x) = 0$ (or ∞) so that $\lim_{x \to a} \frac{f(x)}{g(x)}$ **appears** to lead to "0/0" (or "∞/∞"). If we assume f and g are differentiable in an interval containing $x = a$ (which is a stronger assumption than we actually need), L'Hôpital's Rule states $\lim_{x \to a} \frac{f(x)}{g(x)} = \lim_{x \to a} \frac{f'(x)}{g'(x)}$. Why does this work?

Since f and g are continuous at $x = a$, $\lim_{x \to a} f(x) = \lim_{x \to a} g(x) = 0 = f(a) = g(a)$.

Therefore, $\frac{f(x) - f(a)}{g(x) - g(a)} \overset{f(a)=g(a)=0}{=} \frac{f(x)}{g(x)}$. This is the ratio of **how much** f changes relative to g between a and x. $\frac{f'(x)}{g'(x)}$ measures **how fast** f changes relative to g.

If x is **near** a, and if f changes, for example, about 3 times as fast as g, then f should change about 3 times as much as g. In other words, when x is "close to" a, $\frac{f(x)}{g(x)} \doteq \frac{f'(x)}{g'(x)}$. Take the limit as $x \to a$ and we have L'Hôpy's Rule! Of course, this is not a proof, but I hope it makes the rule plausible. Let's try this rule out.

Example 1) Evaluate $\lim_{x \to 3} \frac{x^2 - 9}{x - 3}$ using L'Hôpital's rule.

Solution

$\lim_{x \to 3} \frac{x^2 - 9}{x - 3}$

$\boxed{\text{"0/0", use L'Hôpital's Rule, take the derivative of the top and bottom.}}$

$= \lim_{x \to 3} \frac{2x}{1} = 6$

Example 2) Evaluate $\lim_{x \to \infty} \frac{x^2}{e^x}$.

Solution

$\lim_{x \to \infty} \frac{x^2}{e^x}$

$\boxed{\text{"∞/∞", use L'Hôpital's Rule, take the derivative of the top and bottom.}}$

$= \lim_{x \to \infty} \frac{2x}{e^x}$

$\boxed{\text{"∞/∞" still! Use L'Hôpital's Rule, again!}}$

$= \lim_{x \to \infty} \frac{2}{e^x} = 0.$

L'Hôpital's Rule Disguised: Converting "Indeterminate Forms" to L'Hôpital Fractions

L'Hôpital's rule gives us a method for dealing with "0/0" or "∞/∞" limits. We can use it on other indeterminate forms* such as $"0^0"$, $"\infty \times 0"$, $"1^\infty"$, $"\infty - \infty"$, etc. We just need to take a few steps to make these forms look like "0/0" or "∞/∞".

Example 1) Evaluate $\lim\limits_{x \to \infty} \left((x^3 + 2x^2 - 1)^{\frac{1}{3}} - x \right)$.

Solution $\lim\limits_{x \to \infty} \left((x^3 + 2x^2 - 1)^{\frac{1}{3}} - x \right) \xlongequal{\text{"}\infty-\infty\text{" Make a fraction!}} \lim\limits_{x \to \infty} \left((x^3 + 2x^2 - 1)^{\frac{1}{3}} - x \right) \left(\dfrac{\frac{1}{x}}{\frac{1}{x}} \right)$

$= \lim\limits_{x \to \infty} \dfrac{\left(\dfrac{(x^3 + 2x^2 - 1)^{\frac{1}{3}}}{x} - 1 \right)}{\frac{1}{x}} \xlongequal{\text{Note } x = (x^3)^{\frac{1}{3}} \text{ so,} \;\; \frac{(x^3+2x^2-1)^{\frac{1}{3}}}{x} = \left(\frac{x^3+2x^2-1}{x^3}\right)^{\frac{1}{3}}} \lim\limits_{x \to \infty} \dfrac{\left(\left(1 + \dfrac{2}{x} - \dfrac{1}{x^3}\right)^{\frac{1}{3}} - 1 \right)}{\frac{1}{x}}$

$\xlongequal{\text{"0/0", Use L'Hôpital's Rule.}} \lim\limits_{x \to \infty} \dfrac{\frac{1}{3}\left(1 + \frac{2}{x} - \frac{1}{x^3}\right)^{-\frac{2}{3}} \left(-\frac{2}{x^2} + \frac{3}{x^4}\right)}{-\frac{1}{x^2}}$

$\xlongequal{\text{Bring the denominator up.}} \lim\limits_{x \to \infty} \left(\dfrac{1}{3}\left(1 + \dfrac{2}{x} - \dfrac{1}{x^3}\right)^{-\frac{2}{3}} \left(-\dfrac{2}{x^2} + \dfrac{3}{x^4}\right)(-x^2) \right)$

$= \lim\limits_{x \to \infty} \left(\dfrac{1}{3}\left(1 + \dfrac{2}{x} - \dfrac{1}{x^3}\right)^{-\frac{2}{3}} \left(2 - \dfrac{3}{x^2}\right) \right) = \left(\dfrac{1}{3}(1 + 0 - 0)^{-\frac{2}{3}}(2 - 0) \right) = \dfrac{2}{3}$

Example 2) Evaluate $\lim\limits_{x \to 0^+} x^x$.

Solution $\lim\limits_{x \to 0^+} x^x \xlongequal{\text{"}0^0\text{". First we use } e^{\ln(*)} = *.} \lim\limits_{x \to 0^+} e^{\ln(x^x)} = \lim\limits_{x \to 0^+} e^{x \ln(x)} \xlongequal{\text{Make a L'Hôpital fraction.}} \lim\limits_{x \to 0^+} e^{\frac{\ln(x)}{\frac{1}{x}}}$

$\xlongequal{\lim\limits_{x \to a} e^{f(x)} = e^{\lim\limits_{x \to a} f(x)}} e^{\left(\lim\limits_{x \to 0^+} \frac{\ln(x)}{\frac{1}{x}} \right)} \xlongequal{\text{"}-\infty/\infty\text{", use L'Hôpital's Rule.}} e^{\left(\lim\limits_{x \to 0^+} \frac{\frac{1}{x}}{-\frac{1}{x^2}} \right)} = e^{\left(\lim\limits_{x \to 0^+} -x \right)} = e^0 = 1$

*For example, $"0^0"$ is an indeterminate form because the answer can be 0, ∞, a contstant, or "does not exist", depending on how the base and the exponent approach zero relative to each other.

Domain (Food for a Function!)

Compare these two problems:
1) Find the domain of the function $y = \sqrt{x-1}$ and then graph the function.
2) Graph the function $y = \sqrt{x-1}$.

While the second question doesn't explicitly ask for the domain, you still need to know it in order to draw the graph.

DOMAIN CATALOG

Expression	Domain
$\sqrt{x}, \sqrt[4]{x}, \sqrt[6]{x}$, etc.	$x \geq 0$
$\dfrac{1}{x}$	$x \neq 0$
$\log_a(x)$, including $\log(x)$ and $\ln(x)$	$x > 0$
$\tan(x), \sec(x)$	$x \neq \pm\dfrac{\pi}{2}, \pm\dfrac{3\pi}{2}, \pm\dfrac{5\pi}{2}, \ldots$
$\cot(x), \csc(x)$	$x \neq \pm\pi, \pm 2\pi, \pm 3\pi, \ldots$
When you learn a new function…	…you will learn the restrictions that come with it.

Example 1) Find the domain for each of the following functions.

(a) $y = \sqrt{x-6}$ (b) $y = \sqrt{9-x^2}$ (c) $y = \dfrac{1}{\sqrt{x^2-9}}$ (d) $y = \sqrt[3]{x-4} = (x-4)^{\frac{1}{3}}$

Solution

[Set Notation] [Interval Notation — Square bracket: **include** the end point. Round bracket: **exclude** the end point.]

(a) $x - 6 \geq 0 \Leftrightarrow x \geq 6$ ∴ Domain $= \{x \in \mathbb{R} \mid x \geq 6\} = [6, \infty)$

(b) $9 - x^2 \geq 0 \Leftrightarrow 9 \geq x^2 \Leftrightarrow x^2 \leq 9 \Leftrightarrow -3 \leq x \leq 3$ ∴ Domain $= [-3, 3]$

(c) $x^2 - 9 > 0 \Leftrightarrow x^2 > 9 \Leftrightarrow |x| > 3 \Leftrightarrow x < -3$ or $x > 3$ ∴ Domain $= (-\infty, -3) \cup (3, \infty)$

(d) There is no restriction. We CAN take the cube root (or the fifth root or the seventh root) of any number, − or +, so we don't need $x - 4 \geq 0$! ∴ Domain $= \mathbb{R}$

Example 2) Find the domain for each: (a) $y = \ln(6 + 5x)$ (b) $y = \tan(2x)$

Solution (a) $6 + 5x > 0 \Leftrightarrow 6 > -5x \Leftrightarrow -\dfrac{6}{5} < x$ ∴ Domain $= \left(-\dfrac{6}{5}, \infty\right)$

[We divided by a "−" and therefore, reversed the inequality!]

(b) $2x \neq \pm\dfrac{\pi}{2}, \pm\dfrac{3\pi}{2}, \pm\dfrac{5\pi}{2}, \ldots$ and so $x \neq \pm\dfrac{\pi}{4}, \pm\dfrac{3\pi}{4}, \pm\dfrac{5\pi}{4}, \ldots$

So, to be concise, Domain $= \left\{x \in \mathbb{R} \mid x \neq \dfrac{\pi}{4} + \dfrac{k\pi}{2}, \text{ where } k \in \mathbb{Z}\right\}$!

Two For You – Variations on $\lim\limits_{\theta \to 0}\left(\dfrac{\sin\theta}{\theta}\right) = 1$

Evaluate the following limits:

1) $\lim\limits_{\theta \to 0}\left(\dfrac{\sin 5\theta}{\sin 3\theta}\right)$

2) $\lim\limits_{h \to 0}\left(\dfrac{\sin(x+h) - \sin x}{h}\right)$ (Hint: use $\sin(x+h) = \sin x \cos h + \cos x \sin h$.)

Answers 1) $\dfrac{5}{3}$

2) $\cos x$ (Here, you have just shown the derivative of $\sin x$ is $\cos x$!)

Three For You – L'Hôpital's Rule

Evaluate the following limits using L'Hôpital's rule:

1) $\lim\limits_{x \to 10}\left(\dfrac{x^2 - 100}{x - 10}\right)$ 2) $\lim\limits_{x \to 0}\left(\dfrac{e^x - x - 1}{x^2}\right)$ 3) $\lim\limits_{x \to 0} x \cot(x)$ (Hint: $\cot(x) = \dfrac{1}{\tan(x)}$)

Answers 1) 20 2) $\dfrac{1}{2}$ 3) 1

Two For You – L'Hôpital's Rule Disguised

Evaluate the following limits using L'Hôpital's rule:

1) $\lim\limits_{x \to \infty} \left((x^3 + 2x - 1)^{\frac{1}{3}} - x \right)$

2) $\lim\limits_{x \to \infty} (e^x + x)^{\frac{2}{x}}$ (Hint: use $\lim\limits_{x \to \infty} (e^x + x)^{\frac{2}{x}} = \lim\limits_{x \to \infty} e^{\ln\left((e^x + x)^{\frac{2}{x}}\right)}$.)

Answers 1) 0 2) e^2

Two For You – Domain (Food for a Function!)

1) State the domain and range of $f(x) = \sqrt{x+1} + 1$ and draw the graph.

2) Write $y = \sqrt{(x-4)^2}$ using first absolute value and then a "branch" definition.

Answers

1) Domain $= [-1, \infty)$, Range $= [1, \infty)$

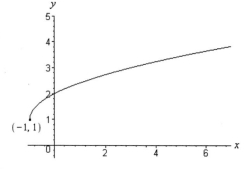

2) $y = |x - 4| = \begin{cases} -(x-4), & \text{if } x < 4 \\ x - 4, & \text{if } x \geq 4 \end{cases} = \begin{cases} 4 - x, & \text{if } x < 4 \\ x - 4, & \text{if } x \geq 4 \end{cases}$

Composite Functions

Given two functions $y = f(x)$ and $y = g(x)$, we can add, subtract, multiply, or divide them to create new functions. For example, $(f+g)(x) \underset{\text{This is what we define } f+g \text{ to be.}}{=} f(x) + g(x)$.

Note: Domain$(f+g)$ = Domain$(f) \cap$ Domain(g). The same is true for $f-g$, $f \cdot g$, and $\dfrac{f}{g}$, with the exception that the domain of $\dfrac{f}{g}$ excludes x values where $g(x) = 0$.

Another common method of combining f and g is to form the composite function, denoted $y = f \circ g(x)$. Here we input x into g, find the output $g(x)$ and plug **this output** into f, that is, $f \circ g(x) = f(g(x))$. Not too hard. Domain is a little more tricky.

Example 1) Let $f(x) = \sqrt{x+3}$ and $g(x) = \dfrac{1}{x}$.

(a) Find $f \circ g(x)$ and determine its domain.   (b) Find $g \circ f(x)$ and determine its domain.

Solution Note Domain$(f) \underset{\text{using set notation}}{=} \{x \in \mathbb{R} \mid x \geq -3\}$ and Domain$(g) = \{x \in \mathbb{R} \mid x \neq 0\}$.

(a) $f \circ g(x) = f(g(x)) = f\left(\dfrac{1}{x}\right) = \sqrt{\dfrac{1}{x} + 3}$

For $f \circ g$, we need $x \in$ Domain(g) and $g(x) = \dfrac{1}{x} \in$ Domain(f).

So $x \neq 0$ and $\dfrac{1}{x} \geq -3$. Now, $\dfrac{1}{x} \geq -3 \Leftrightarrow \dfrac{1}{x} + 3 \geq 0 \Leftrightarrow \dfrac{1 + 3x}{x} \geq 0 \Leftrightarrow \dfrac{3\left(x + \dfrac{1}{3}\right)}{x} \geq 0$

After an easy number line analysis, we find $x \leq -\dfrac{1}{3}$ or $x > 0$.

$\therefore$ Domain$(f \circ g) = \{x \in \mathbb{R} \mid x \leq -\dfrac{1}{3}$ or $x > 0\} \underset{\text{interval notation}}{=} (-\infty, -\dfrac{1}{3}] \cup (0, \infty)$

(b) $g \circ f(x) = g(f(x)) = g(\sqrt{x+3}) = \dfrac{1}{\sqrt{x+3}}$

For $g \circ f$, we need $x \in$ Domain(f) and $f(x) = \sqrt{x+3} \in$ Domain(g). So, $x \geq -3$ and $\sqrt{x+3} \neq 0$. Now, $\sqrt{x+3} \neq 0 \Leftrightarrow x \neq -3$

$\therefore$ Domain$(g \circ f) = \{x \in \mathbb{R} \mid x > -3\} \underset{\text{interval notation}}{=} (-3, \infty)$

Continuity and Discontinuity at a Point

f is continuous at $x = a$ if $\lim_{x \to a} f(x) = f(a)$.

Continuous functions have no holes or breaks or vertical asymptotes. You have to have a **limit**, you have to have a **function value**, and they must be **equal**.

Example 1) Explain why each function is discontinuous at the indicated value:

(a) $f(x) = \dfrac{x^2 - 16}{x - 4}$ at $x = 4$ (b) $f(x) = \dfrac{x}{|x|}$ at $x = 0$

(c) $f(x) = \begin{cases} x^2, & \text{if } x \neq 0 \\ 2, & \text{if } x = 0 \end{cases}$ at $x = 0$ (d) $f(x) = [[x]]$ at $x = 2$

Solution (a) $\lim_{x \to 4} f(x) = \lim_{x \to 4} \dfrac{(x-4)(x+4)}{x-4} = 8$,

so the limit exists, but $f(4)$ does not exist.

(b) $f(x) = \dfrac{x}{|x|} = \begin{cases} -1, & \text{if } x < 0 \\ 1, & \text{if } x > 0 \end{cases}$

$\therefore \lim_{x \to 0^-} f(x) = \lim_{x \to 0^-}(-1) = -1$ and $\lim_{x \to 0^+} f(x) = \lim_{x \to 0^+} 1 = 1$ and so $\lim_{x \to 0} f(x)$ does not exist.

(c) $\lim_{x \to 0} f(x) = \lim_{x \to 0} x^2 = 0$ so the limit exists. But $f(0) = 2 \neq \lim_{x \to 0} f(x)$.

(d) $\lim_{x \to 2^-} f(x) \overset{\text{For } 1 \leq x < 2,\ f(x) = 1.}{=} \lim_{x \to 2^-}(1) = 1$ and $\lim_{x \to 2^+} f(x) = \lim_{x \to 2^+} 2 = 2$,

and so $\lim_{x \to 2} f(x)$ does not exist.

Note that $f(2) = 2$ and so f is continuous from the **right**.

Example 2) For the function at the right, give a reason why it is discontinuous at each of a, b, and c. State whether it is continuous from the left or right.

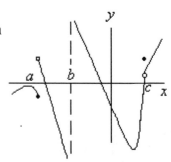

Solution The function is discontinuous at a and b because the limit does not exist. It is discontinuous at c because the limit and value are unequal. The function is continuous from the **left** at a but from neither side at b and c.

Continuous Functions (Intervals of Continuity)

Continuous functions have no holes. When you draw a continuous function, your pen never leaves the paper. When is a function **not continuous**? Basically, you have to check **domain restrictions**, **division by 0**, and **"branches"** (which we look at geometrically in Example 2 and using function definitions in the next section!) When you add, subtract, and multiply continuous functions, you get continuous functions. Same for division, as long as you don't have 0 in the bottom!

Example 1)(a) Based on your mathematical experience, which of the following are continuous: polynomials, trigonometric functions, $y = a^x$, $y = \log_a x$?
(b) If f and g are continuous, discuss the continuity of $f+g$, $f-g$, fg, f/g.

Solution
(a) Polynomials, $y = \sin x$, $y = \cos x$, and $y = a^x$ are continuous for $x \in \mathbb{R}$.
$y = \log_a x$ is continuous for $x > 0$.
$y = \tan x$ and $y = \sec x$ are continuous for $x \neq \pi/2 + k\pi$, where $k \in \mathbb{Z}$.
$y = \cot x$ and $y = \csc x$ are continuous for $x \neq \pi + k\pi$, where $k \in \mathbb{Z}$.
(b) $f+g$, $f-g$, and fg are all continuous whenever f and g are.
f/g is continuous as long as f and g are and $g(x) \neq 0$.

Example 2) State the intervals of continuity for this function....... →

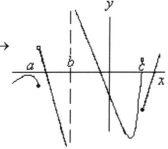

Solution The function is continuous on the intervals $(-\infty, a]$, (a,b), (b,c), and $[c, \infty)$.
Note that the function is continuous from the **left** at $x = a$ and from the **right** at $x = c$.

Example 3) State the intervals on which the following are continuous:

(a) $f(x) = \sqrt{x-3}$ (b) $g(x) = \ln(|x|)$ (c) $h(x) = \dfrac{x^{1/4}}{(x-5)(x+3)}$

Solution
(a) Since $x - 3 \geq 0$, therefore $x \geq 3$ and so f is continuous on the interval $[3, \infty)$.
(b) Since $|x| \geq 0$, we only have to exclude $x = 0$ and so g is continuous on $(-\infty, 0)$ and $(0, \infty)$.
(c) $x \neq 5$ or -3. But $x^{1/4}$ is only defined for $x \geq 0$, so -3 is already gone. Therefore h is continuous on $[0,5)$ and $(5, \infty)$.

Continuity and Branch Functions

There are two kinds of branch functions: explicit and implicit. Apart from the usual continuity concerns (division by 0, positive domain for logs, etc.) you have to be careful **at the values of x where the function branches.** Remember:

$$f \text{ is continuous at } x = a \text{ if } \lim_{x \to a} f(x) = f(a).$$

Example 1) Discuss the continuity of the function $f(x) = \begin{cases} x, & \text{if } x < 0 \\ 2, & \text{if } x = 0 \\ x^2, & \text{if } 0 < x < 2 \\ x+2, & \text{if } x \geq 2 \end{cases}$

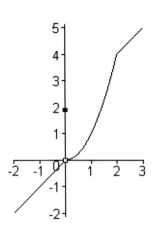

Solution The only possible problems are at $x = 0$ and $x = 2$.

$\lim_{x \to 0^-} f(x) = \lim_{x \to 0^-} x = 0 \qquad \lim_{x \to 0^+} f(x) = \lim_{x \to 0^+} x^2 = 0$

Since the left- and right-hand limits are equal, $\lim_{x \to 0} f(x) = 0$.

But $f(0) = 2$ and so f **is not** continuous at $x = 0$.

$\lim_{x \to 2^-} f(x) = \lim_{x \to 2^-} x^2 = 4 \qquad \lim_{x \to 2^+} f(x) = \lim_{x \to 2^+} (x+2) = 4.$

The left-hand and right-hand limits are equal, so $\lim_{x \to 2} f(x) = 4$. Also, $f(2) = 4$ and so f **is** continuous at $x = 2$. f is continuous on the intervals $(-\infty, 0)$ and $(0, \infty)$.

Example 2) Discuss the continuity of the function $g(x) = \dfrac{|x|}{x}$.

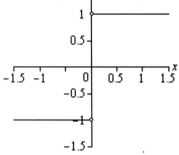

Solution The only possible problem here is at $x = 0$.
Writing g explicitly as a branch function,

we have $g(x) = \begin{cases} \dfrac{-x}{x}, & \text{if } x < 0 \\ \dfrac{x}{x}, & \text{if } x > 0 \end{cases} = \begin{cases} -1, & \text{if } x < 0 \\ 1, & \text{if } x > 0 \end{cases}$

$\lim_{x \to 0^-} g(x) = \lim_{x \to 0^-} (-1) = -1$ and $\lim_{x \to 0^+} g(x) = \lim_{x \to 0^+} 1 = 1$

so $\lim_{x \to 0} g(x)$ does not exist. Therefore, g is **not continuous** at $x = 0$ and so g is continuous on the intervals $(-\infty, 0)$ and $(0, \infty)$.

Two For You – Composite Functions

1) Let $f(x) = \sqrt{x}$ and $g(x) = \dfrac{1}{x^2 - 1}$.

(a) Find $f \circ g(x)$ and find its domain. (b) Find $g \circ f(x)$ and find its domain.

2) (a) What is the domain of $h(x) = \dfrac{1}{x-1}$?

(b) In question 1(b), you found $g \circ f(x) = \dfrac{1}{x-1}$ but its domain is different from h. Why?

Answer 1(a) $f \circ g(x) = \dfrac{1}{\sqrt{x^2 - 1}}$ Domain$(f \circ g)$ [using interval notation] $= (-\infty, -1) \cup (1, \infty)$

(b) $g \circ f(x) = \dfrac{1}{x-1}$ Domain$(g \circ f)$ [using interval notation] $= [0, 1) \cup (1, \infty)$

2(a) Domain$(h) = (-\infty, 1) \cup (1, \infty)$

(b) In 1(b), we must have $x \in$ Domain$(f) = [0, \infty)$!

Two For You – Continuity and Discontinuity at a Point

For each of the following, give a reason why the function is not continuous at the indicated value:

1) $f(x) = \dfrac{1}{x-3}$ at $x = 3$ 2) $f(x) = \begin{cases} x+1, & \text{if } x < -2 \\ 2, & \text{if } x = -2 \\ x^2 - 5, & \text{if } x > -2 \end{cases}$ at $x = -2$

Answers 1) $\lim\limits_{x \to 3} f(x)$ does not exist **OR** $f(3)$ does not exist (either answer).

2) $\lim\limits_{x \to -2} f(x) = -1 \neq f(-2)$

Two For You – Continuous Functions (Intervals of Continuity)

State the intervals of continuity for f and g.

1) $f(x) = \dfrac{\sqrt{3x-9}}{x-5}$

2) $g(x) = [[x]]$, where $-2 \leq x \leq 2$

(g is "the greatest integer less than or equal to x" or "floor" function.)

Answers 1) $[3,5), (5,\infty)$ 2) $[-2,-1), [-1,0), [0,1), [1,2)$

Two For You – Continuity and Branch Functions

State the intervals of continuity for f and g.

1) $f(x) = \begin{cases} \sin x, & \text{if } x < \pi \\ \cos x, & \text{if } x \geq \pi \end{cases}$ 2) $g(x) = \dfrac{x+3}{|x+3|}$

Answers 1) $(-\infty, \pi), [\pi, \infty)$ 2) $(-\infty, -3), (-3, \infty)$

Essential versus Removable Discontinuities

The discontinuity is **removable** if it is a "hole" in the function. In this case, there is a limit and there is a function value, and they are different **or** there is no function value.

The discontinuity is **essential** if it is **any other type of discontinuity**. In this case, there is definitely **NO** limit, though this can happen for a variety of reasons.

Example 1) Each of the following functions has a discontinuity at one point. Classify it as either essential or removable. If essential, give the reason. If removable, redefine the function to make it continuous at the point.

(a) $f(x) = \begin{cases} \sin x, & \text{if } x < 0 \\ 1, & \text{if } x = 0 \\ x, & \text{if } x > 0 \end{cases}$
(b) $g(x) = \dfrac{x^2 - 9}{x - 3}$
(c) $h(x) = \dfrac{1}{x^2}$

(d) $j(x) = \begin{cases} x, & \text{if } x < 0 \\ 1, & \text{if } x = 0 \\ x+1, & \text{if } x > 0 \end{cases}$
(e) $k(x) = \sin\left(\dfrac{1}{x}\right)$

Solution

(a) Removable. Define $f(0) = 0$.

(b) Removable. The problem is at $x = 3$:

$\lim\limits_{x \to 3} g(x) = \lim\limits_{x \to 3} \dfrac{(x-3)(x+3)}{x-3} = 6$, but $g(3)$ is not defined. Define $g(3) = 6$.

(c) Essential. $\lim\limits_{x \to 0} h(x) = \infty$

(d) Essential. $\lim\limits_{x \to 0^-} j(x) = \lim\limits_{x \to 0^-} x = 0$ while $\lim\limits_{x \to 0^+} j(x) = \lim\limits_{x \to 0^+} (x+1) = 1$
and so $\lim\limits_{x \to 0} j(x)$ does not exist.

(e) Essential. $\lim\limits_{x \to 0} k(x)$ does not exist. In fact, $\sin\left(\dfrac{1}{x}\right)$ oscillates between -1 and 1 **faster** and **faster** as x gets closer and closer to 0.

Finding the Derivative from the Definition

$$f \text{ is differentiable at } x = a \text{ if } \lim_{h \to 0} \frac{f(a+h) - f(a)}{h} = f'(a).$$

The alternate form of this definition is

$$f \text{ is differentiable at } x = a \text{ if } \lim_{x \to a} \frac{f(x) - f(a)}{x - a} = f'(a).$$

The second definition is the same as the first. Many of you find this hard to believe. The **variable** in the second formulation is $x = a + h$ and so $x - a = h$. As h approaches 0, x approaches a. So, in the first definition, replace:

$$h \to 0 \text{ with } x \to a \qquad h \text{ with } x - a \qquad a + h \text{ with } x$$

Lo and behold (not by magic!), you have the second definition. Why do some students confuse these? I think it is because teachers ask you to find $f'(\boxed{x})$, rather than $f'(\boxed{a})$, from the definition when you first study derivatives. This is okay in the first definition where x takes on the role of a. **BUT** if you try to use the second definition, you would be forcing x to be both $(a+h)$ and a at the same time!

Example 1) Find the derivative for the function $f(x) = x^2 + x + 1$ at $(a, f(a))$
(a) using the first definition   (b) using the second definition.

Solution (a) $f'(a) = \lim\limits_{h \to 0} \dfrac{f(a+h) - f(a)}{h} = \lim\limits_{h \to 0} \dfrac{(a+h)^2 + (a+h) + 1 - (a^2 + a + 1)}{h}$

$= \lim\limits_{h \to 0} \dfrac{a^2 + 2ah + h^2 + a + h + 1 - a^2 - a - 1}{h} = \lim\limits_{h \to 0} \dfrac{2ah + h^2 + h}{h} = \lim\limits_{h \to 0} \dfrac{\cancel{h}(2a + h + 1)}{\cancel{h}}$

$= 2a + 1$

(b) $f'(a) = \lim\limits_{x \to a} \dfrac{f(x) - f(a)}{x - a} = \lim\limits_{x \to a} \dfrac{x^2 + x + 1 - (a^2 + a + 1)}{x - a}$

$= \lim\limits_{x \to a} \dfrac{x^2 - a^2 + x - a}{x - a} = \lim\limits_{x \to a} \dfrac{(x-a)(x+a+1)}{x - a} = \lim\limits_{x \to a} \dfrac{\cancel{(x-a)}(x+a+1)}{\cancel{x-a}} = 2a + 1$

Differentiable Functions (Intervals of Differentiability)

When is a **continuous** function **not differentiable**?

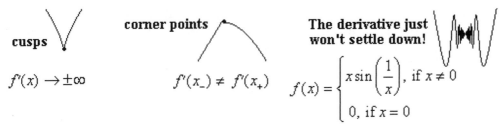

Also, check **domain restrictions**, **division by 0 in the derivative**, and "**branches**". When you add, subtract and multiply differentiable functions, you get differentiable functions. Same for division, as long as you don't have 0 in the bottom!

Example 1) Based on your mathematical experience, which of the following are differentiable: polynomials, trigonometric functions, $y = a^x$, $y = \log_a x$, $y = |x|$?

Solution Polynomials, $y = \sin x$, $y = \cos x$, and $y = a^x$ are differentiable for $x \in \mathbb{R}$.
$y = \log_a x$ is differentiable for $x > 0$.
$y = \tan x$ and $y = \sec x$ are differentiable for $x \neq \pi/2 + k\pi$, where $k \in \mathbb{Z}$.
$y = \cot x$ and $y = \csc x$ are differentiable for $x \neq \pi + k\pi$, where $k \in \mathbb{Z}$.

$y = |x|$ has a $\boxed{\text{corner point at } x = 0}$ and so the function is differentiable for $x \neq 0$.
(Corner Point at $x=0$: $f'(0_-) \neq f'(0_+)$)

Example 2) State the intervals of differentiability for this function... →

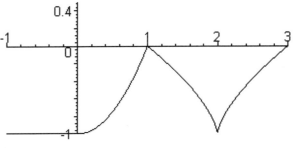

Solution The function is differentiable on the intervals $[-1, 1]$, $[1, 2)$, and $(2, 3]$. Note that $x = 1$ is a corner point and $x = 2$ is a cusp point.

Example 3) State the intervals on which the following are differentiable.

(a) $f(x) = \sqrt{x-3}$ (b) $h(x) = \dfrac{x^{1/4}}{(x-5)(x+3)}$

Solution
(a) Since $x - 3 \geq 0$, therefore $x \geq 3$ and so f is differentiable on the interval $[3, \infty)$.
(b) $x \neq 5$ or -3. But $x^{1/4}$ is only defined for $x \geq 0$, so -3 is already gone! Therefore, h is differentiable on the intervals $[0, 5)$ and $(5, \infty)$.

Differentiability and Branch Functions

There are two kinds of branch functions: explicit and implicit. Apart from the usual differentiability concerns (division by 0 in the derivative, positive domain for logs, etc.) you have to be careful **at the values of x where the function branches. Also, we will deal here only with functions that are continuous on their domains.**

Example 1) Discuss the differentiability of the function $f(x) = \begin{cases} x, & \text{if } x < 0 \\ x^2, & \text{if } 0 \leq x < 2 \\ 4x - 4, & \text{if } x \geq 2 \end{cases}$

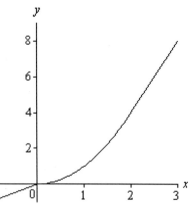

Solution The only possible problems are at $x = 0$ and $x = 2$.

$x = 0$: $\lim_{x \to 0^-} f'(x) = \lim_{x \to 0^-} 1 = 1 \quad \lim_{x \to 0^+} f'(x) = \lim_{x \to 0^+} 2x = 0$

Since the left- and right-hand limits are unequal, $f'(0)$ does not exist.

> **Note:** f **is** differentiable from both the left and the right at $x = 0$ and yet $f'(0)$ does not exist.

$x = 2$: $\lim_{x \to 2^-} f'(x) = \lim_{x \to 2^-} 2x = 4 \quad \lim_{x \to 2^+} f'(x) = \lim_{x \to 2^+} 4 = 4.$

Therefore, $f'(2) = 4$ and so f is differentiable for $x \neq 0$.

> **Note continued:** If a function is discontinuous at a value of x, it can be continuous from the left or the right or neither, but not both. As this example shows, a continuous function may not be differentiable at a value of x yet still be differentiable from the left **and** the right.

Example 2) Discuss the differentiability of the function $g(x) = |x^3|$.

Solution The only possible problem here is at $x = 0$.
Writing g explicity as a branch function,
we have $g(x) = \begin{cases} -x^3, & \text{if } x < 0 \\ x^3, & \text{if } x \geq 0 \end{cases}$

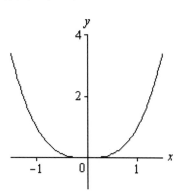

$\lim_{x \to 0^-} g'(x) = \lim_{x \to 0^-}(-3x^2) = 0$ and $\lim_{x \to 0^+} g(x) = \lim_{x \to 0^+} 3x^2 = 0$
so $g''(0) = 0$ and g is differentiable for $x \in \mathbb{R}$.

Two For You – Essential versus Removable Discontinuities

Each function has a single discontinuity. State the problem x value and whether the discontinuity is essential or removable. If essential, state why. If removable, redefine the function to make it continuous.

1) $f(x) = \dfrac{|x-5|}{x-5}$ 2) $g(x) = \dfrac{\sin x}{x}$

Answers 1) Essential at $x = 5$ since $\lim\limits_{x \to 5} f(x)$ does not exist.
2) Removable at $\theta = 0$. Define $g(0) = 1$.

Two For You – Finding the Derivative from the Definition

Set up the first step to find $f'(a)$ for $f(x) = \dfrac{1}{\sqrt{x}}$

1) using the first definition   2) using the second definition.

Answers 1) $f'(a) = \lim\limits_{h \to 0} \dfrac{\dfrac{1}{\sqrt{a+h}} - \dfrac{1}{\sqrt{a}}}{h}$ 2) $f'(a) = \lim\limits_{x \to a} \dfrac{\dfrac{1}{\sqrt{x}} - \dfrac{1}{\sqrt{a}}}{x - a}$

Two For You – Differentiable Functions

State the intervals of differentiability:

1) $f(x) = \dfrac{\sqrt{3x-9}}{x-5}$

2) $g(x) = [[x]]$, if $-2 \leq x \leq 2$

Answers 1) $[3, 5), (5, \infty)$ 2) $[-2,-1), [-1, 0), [0, 1), [1, 2)$

Two For You – Differentiability and Branch Functions

State the intervals of differentiability:

1) $f(x) = \begin{cases} \sin x - \pi, & \text{if } x < \pi \\ \cos x - x + 1, & \text{if } x \geq \pi \end{cases}$

2) $g(x) = |x+3|$

Answers 1) $(-\infty, \infty)$ 2) $(-\infty, -3], [-3, \infty)$

Critical Numbers

A **critical number** of a function $y = f(x)$ is a number c in the domain of f where (i) $f'(c) = 0$ or (ii) $f'(c)$ does not exist or (iii) $(c, f(c))$ is an end point. We all know what 0 derivatives and end points look like. If $f(x)$ is **continuous** at c but $f'(c)$ does not exist, then one of three things must be happening:

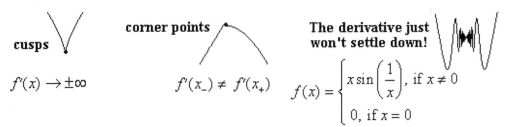

cusps
$f'(x) \to \pm\infty$

corner points
$f'(x_-) \neq f'(x_+)$

The derivative just won't settle down!
$f(x) = \begin{cases} x \sin\left(\dfrac{1}{x}\right), & \text{if } x \neq 0 \\ 0, & \text{if } x = 0 \end{cases}$

A maximum or minimum point of a continuous function f **must occur** at a critical number of f!

The reverse is FALSE! Lots of critical points are neither maxima nor minima!

Example 1) Find all critical numbers of the function $f(x) = 3x^4 + 2x^3$, $-2 \leq x \leq 2$ and classify the corresponding critical points.

Solution $f'(x) = 12x^3 + 6x^2 = 12x^2\left(x + \dfrac{1}{2}\right)$

Both f and f' have domain $[-2, 2]$. The critical numbers are -2, $-\dfrac{1}{2}$, 0, and 2.

$f'(x)$
```
     -2    -    -1/2  +   0       +      2
      |     \    |   /   |       /       |
```

$(-2, 32)$ is an end point maximum;
$(-1/2, -1/16)$ is a minimum where $f' = 0$;
$(0, 0)$ is neither a max nor a min. In fact, it is a point of inflection;
$(2, 64)$ is an end point maximum.

Example 2) Find all critical numbers of the function $f(x) = 5x^{2/3} - x^{5/3}$.

Solution $f'(x) = \dfrac{10}{3}x^{-1/3} - \dfrac{5}{3}x^{2/3} = -\dfrac{5}{3}\left(\dfrac{x-2}{x^{1/3}}\right)$

The critical numbers are 0 and 2. Note that $0 \in \text{domain}(f)$ but not $\text{domain}(f')$.

Max and Min Points from the First Derivative

There are **four** ways a **continuous** function can have a maximum or minimum point:
1) $f'(x) = 0$ 2) $f'(x) \to \pm\infty$ 3) $f'(x_-) \neq f'(x_+)$ 4) an end point
              (a cusp)   (a corner point)

minimum points

end point	$f'(x) = 0$	$f'(x) \to \pm\infty$ cusp	$f'(x_-) \neq f'(x_+)$ corner	end point
$y' > 0$	$y' < 0 \quad y' > 0$	$y' < 0 \quad y' > 0$	$y' < 0 \quad y' > 0$	$y' < 0$
$y' < 0$	$y' > 0 \quad y' < 0$	$y' > 0 \quad y' < 0$	$y' > 0 \quad y' < 0$	$y' > 0$

maximum points

What all minimums have in common: the function decreases ↘ and then increases ↗.

What all maximums have in common: the function increases ↗ and then decreases ↘.

(For end points, "half" of each statement applies!)

Example 1) Given $f'(x) = \dfrac{(x-2)^5(x+2)^3}{(x-1)^{1/3}}$, $-3 \leq x \leq 3$, identify the values of x where $f(x)$ has maximum and minimum points. Classify these extremes as one of the 4 types. (You may assume that $f(x)$ is defined for $x \in [-3, 3]$, including $x = 1$.)

Solution Analyze the sign of $f'(x)$ on $[-3, 3]$. The significant values are $x = -3, -2, 1, 2,$ and 3.

$f'(x)$:
-3 (−)(−)(−) − dec ↘
-2 (+)(−)(−) + inc ↗
1 (+)(+)(−) − dec ↘
2 (+)(+)(+) + inc ↗
3

$x = -3$: end point, $(-3, f(-3))$ is a maximum point.
$x = -2$: $f'(-2) = 0$, $(-2, f(-2))$ is a minimum point.
$x = 1$: $f'(1) \to \pm\infty$, cusp, $(1, f(1))$ is a maximum point.
$x = 2$: $f'(2) = 0$, $(2, f(2))$ is a minimum point.
$x = 3$: end point, $(3, f(3))$ is a maximum point.

Graphing y vs y' vs y''
(Increasing/Decreasing and Concavity)

Here is a function that illustrates:

$y' > 0$ and $y'' > 0$: y is increasing and concave up.
$y' > 0$ and $y'' < 0$: y is increasing and concave down.
$y' < 0$ and $y'' > 0$: y is decreasing and concave up.
$y' < 0$ and $y'' < 0$: y is decreasing and concave down.

Example 1) Insert the letters **a, b, c,** and **d** with the correct choice of y' and y'' in the above graph.

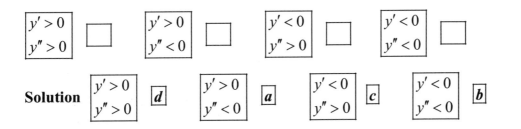

Example 2)(a) By estimating the slopes of the tangents to $y = f(x)$, sketch the graph of $y' = f'(x)$.
(b) By estimating the slopes of the tangents to $y' = f'(x)$, sketch the graph of $y'' = f''(x)$.

Solution m = slope

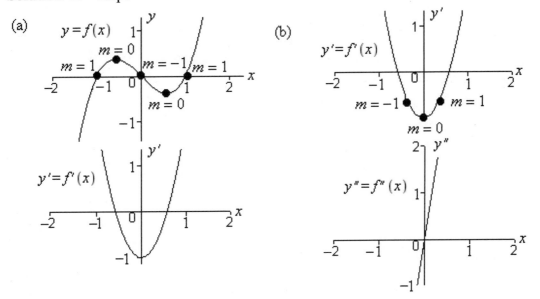

Graph Sketching with Calculus

We know how to find intercepts, intervals of increasing and decreasing, intervals of concave up and concave down, extreme points, and points of inflection. **PLEASE REVIEW THE PREVIOUS THREE SECTIONS AND THEN** let's put it all together to accurately graph a function.

Example 1) The function $y = 3x^4 + 2x^3 = 3x^3(x+2/3)$ has first and second derivatives $y' = 12x^2(x+1/2)$ and $y'' = 36x(x+1/3)$. Using intercepts, intervals of increasing and decreasing, intervals of concave up and concave down, extreme points, and points of inflection, create a beautiful graph!*

*When setting up your number line analysis charts, remember that $(x-a)^n$ changes sign at $x = a$ only if n is **odd**!

Solution x intercepts: $y = 0 \Rightarrow x = 0, \, x = -2/3;$ y intercept: $x = 0 \Rightarrow y = 0$

$\dfrac{dy}{dx}$: number line with critical points at $-\dfrac{1}{2}$ (min, $y'=0$) and 0 ($y'=0$); sign: $-$ on $(-\infty, -\dfrac{1}{2})$ dec., $+$ on $(-\dfrac{1}{2}, 0)$ inc., $+$ on $(0, \infty)$ inc. Point $\left(-\dfrac{1}{2}, -\dfrac{1}{16}\right)$, $(0,0)$.

$x = 0 \Rightarrow y = 0$

$x = -\dfrac{1}{2} \Rightarrow y = -\dfrac{1}{16}$

$\dfrac{d^2y}{dx^2}$: number line with points at $-\dfrac{1}{3}$ ($y''=0$, pt. of inflection) and 0 ($y''=0$, pt. of inflection); sign: $+$ C.U. on $(-\infty, -\dfrac{1}{3})$, $-$ C.D. on $(-\dfrac{1}{3}, 0)$, $+$ C.U. on $(0, \infty)$. Point $\left(-\dfrac{1}{3}, -\dfrac{1}{27}\right)$, $(0,0)$. min

$x = -\dfrac{1}{3} \Rightarrow y = -\dfrac{1}{27}$

Summary Chart

Interval	y'	y''	Shape
$\left(-\infty, -\dfrac{1}{2}\right)$	$-$	$+$	╲
$\left(-\dfrac{1}{2}, -\dfrac{1}{3}\right)$	$+$	$+$	╱
$\left(-\dfrac{1}{3}, 0\right)$	$+$	$-$	╱
$(0, \infty)$	$+$	$+$	╱

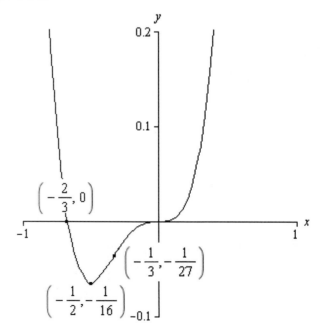

Two For You – Critical Numbers

Find the critical numbers for each of the following functions:

1) $f(x) = \dfrac{x^2}{x^2 - 4}$

2) $g(x) = (x+1)^{1/3}(x-1)^{2/3}$ $\left(\text{Hint: } g'(x) = \dfrac{x + 1/3}{(x+1)^{2/3}(x-1)^{1/3}}\right)$

Answers 1) $x = 0$ 2) $x = -\dfrac{1}{3}, \; x = -1, \; x = 1$

Two For You – Max and Min Points from the First Derivative

For the following, identify the values of x where the function has maxima or minima and classify each extreme:

1) $f'(x) = -\dfrac{x-2}{x^{1/3}}$, where $x \geq -1$ (Hint: be careful with that extra "–"!)

2) $g'(x) = \begin{cases} -x, & \text{if } x < 0 \\ x^2, & \text{if } x \geq 0 \end{cases}$

Answers 1) $x = -1$: end point, maximum; $x = 0$, cusp, minimum; $x = 2$, $f'(x) = 0$, maximum
2) $x = 0$, corner point.

Two For You – Graphing y vs y' vs y''

Match the shapes with the appropriate pair of derivatives as in Example 1 (pg. 179).

1) 2)

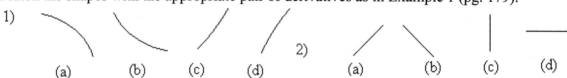

(a)   (b)   (c)   (d)   (a)   (b)   (c)   (d)

Answers 1)(a) $\begin{array}{l} y' < 0 \\ y'' < 0 \end{array}$ (b) $\begin{array}{l} y' < 0 \\ y'' > 0 \end{array}$ (c) $\begin{array}{l} y' > 0 \\ y'' > 0 \end{array}$ (d) $\begin{array}{l} y' > 0 \\ y'' < 0 \end{array}$

2)(a) $\begin{array}{l} y' > 0 \\ y'' = 0 \end{array}$ (b) $\begin{array}{l} y' < 0 \\ y'' = 0 \end{array}$ (c) y', y'' both undefined (d) $y' = y'' = 0$

Two For You – Graph Sketching With Calculus

1) The function $y = x^3 - 9x$ satisfies $y' = 3x^2 - 9 = 3(x - \sqrt{3})(x + \sqrt{3})$ and $y'' = 6x$. Using intercepts, intervals of increasing and decreasing, intervals of concave up and concave down, extreme points, and points of inflection, create a beautiful graph!

2) The function $y = \dfrac{x-1}{x^2}$ satisfies $y' = \dfrac{-(x-2)}{x^3}$ and $y'' = \dfrac{2(x-3)}{x^4}$. Using intercepts, intervals of increasing and decreasing, intervals of concave up and concave down, extreme points, and points of inflection, create a beautiful graph! **Look for horizontal and vertical asymptotes too!**

Answers

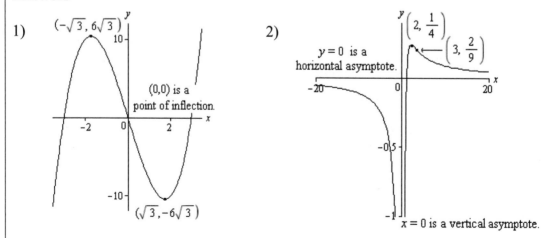

Graph Sketching with Calculus: Vertical Tangent!

FIRST, PLEASE REVIEW THE PREVIOUS FOUR SECTIONS. Let's do a graph sketching example where a **vertical tangent** makes an appearance.

Example 1) The function $y = 5x^{\frac{2}{3}} - x^{\frac{5}{3}} = -x^{\frac{2}{3}}(x-5)$ has first and second derivatives $y' = -\dfrac{5}{3}\left(\dfrac{x-2}{x^{\frac{1}{3}}}\right)$ and $y'' = -\dfrac{10}{9}\left(\dfrac{x+1}{x^{\frac{4}{3}}}\right)$. Using intercepts, intervals of increasing and decreasing, intervals of concave up and concave down, extreme points, and points of inflection, create a beautiful graph!*

*We are dealing with **third roots** here. When setting up your number line analysis charts, remember that $(x-a)^{\frac{n}{3}}$ changes sign at $x = a$ only if n is odd!

Solution x intercepts: $y = 0 \Rightarrow x = 0, \; x = 5;$ $\quad y$ intercept: $x = 0 \Rightarrow y = 0$

$\dfrac{dy}{dx}$: number line with critical points at $(0,0)$ where y' is undefined and $\left(2, 3 \cdot 2^{\frac{2}{3}}\right)$ where $y' = 0$ (max). Sign: $-$ (dec.) for $x<0$, $+$ (inc.) for $0<x<2$, $-$ (dec.) for $x>2$.

$x = 0 \Rightarrow y = 0$ (Note: There is a vertical tangent at $x = 0$.)

$x = 2 \Rightarrow y = -(2)^{\frac{2}{3}}(2-5) = 3(2)^{\frac{2}{3}} \doteq 4.8$

$\dfrac{d^2y}{dx^2}$: number line with points at $(-1, 6)$ where $y'' = 0$ (pt. of inflection) and $(0, 0)$ where y'' is undefined. Sign: $+$ C.U. for $x<-1$, $-$ C.D. for $-1<x<0$, $-$ C.D. for $x>0$.

$x = -1 \Rightarrow y = -(-1)^{\frac{2}{3}}(-1-5) = 6\left((-1)^{\frac{1}{3}}\right)^2 = 6$

Summary Chart

Interval	y'	y''	Shape
$(-\infty, -1)$	$-$	$+$	↘ (concave up)
$(-1, 0)$	$-$	$-$	↘ (concave down)
$(0, 2)$	$+$	$-$	↗ (concave down)
$(2, \infty)$	$-$	$-$	↘ (concave down)

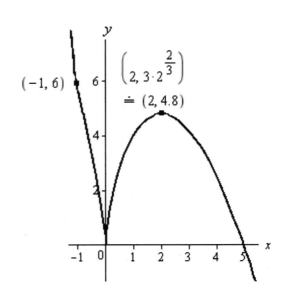

Estimating Using the Differential

Let $y = f(x)$. $\dfrac{dy}{dx} = f'(x)$ gives the slope of the tangent at the point $(x, f(x))$.

If we treat $dx (\neq 0)$ as the run and dy as the rise so that $\dfrac{dy}{dx} = f'(x) =$ slope of the tangent, then $\boxed{dy \underset{\text{Cross multiply by } dx!}{=} f'(x)\,dx}$; in words, **the rise = slope times the run.**

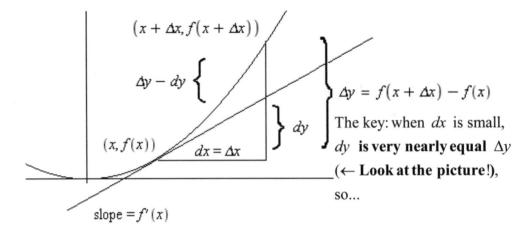

$\Delta y = f(x + \Delta x) - f(x)$

The key: when dx is small, dy **is very nearly equal** Δy (← Look at the picture!), so...

$\boxed{\text{...the new } y = f(x + \Delta x) = \text{old } y + \Delta y \underset{\text{when } dx \text{ is small!}}{\doteq} \text{old } y + dy = f(x) + f'(x)\,dx}$

Example 1) Estimate $\sqrt{4.02}$ using the differential.

Solution We need to choose an appropriate function $f(x)$, a value of x close to 4.02 at which we can **easily** evaluate the function and a **small** dx value.

Let $f(x) = \sqrt{x}$, $x = 4$, and $dx = 0.02$. Then $\dfrac{dy}{dx} = f'(x) = \dfrac{1}{2}x^{-1/2} = \dfrac{1}{2\sqrt{x}}$.

$\therefore dy = f'(x)\,dx = \dfrac{dx}{2\sqrt{x}}$. In this example, $dy \underset{\substack{x=4 \\ dx=0.02}}{=} \dfrac{0.02}{2\sqrt{4}} = \dfrac{1}{200} = 0.005$

$\therefore \sqrt{4.02} \underset{\text{exactly}}{=} \sqrt{4} + \Delta y \doteq \sqrt{4} + dy = 2.005$.

Compare $\sqrt{4.02} \underset{\text{calculator}}{\doteq} 2.00499376558$.

Rolle's Theorem

Suppose $y = f(x)$ is
(i) continuous on $[a, b]$
(ii) differentiable on (a, b) and
(iii) $f(a) = f(b)$.
Then there is **at least one** value
$c \in (a,b)$ such that $f'(c) = 0$.

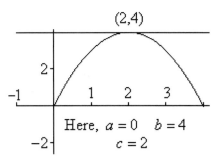

Here, $a = 0$, $b = 4$, $c = 2$

So, basically, Rolle's says that if a continuous function goes, for example, up and then turns around and comes back down to its starting value, then it must have a maximum, say at $x = c$. Since the function is differentiable, this max can't be a corner point nor a cusp. Since it certainly isn't an end point, the only other choice is $f'(c) = 0$.

Example 1) Verify Rolle's Theorem with $f(x) = x^2 - 3$, $a = -1$, and $b = 1$.

Solution f is certainly differentiable and continuous on $(-1,1)$ and $[-1,1]$, respectively. Also, $f(-1) = f(1) = -2$. Rolle's Theorem guarantees $c \in (-1,1)$ satisfying $f'(c) = 0$. Since $f'(x) = 2x$, solving $2c = 0$ gives $c = 0 \in (-1,1)$.

Example 2) Sketch the graph of a function showing
(a) how the conclusion of Rolle's Theorem **can fail** if
  (i) f is not continuous at a
  (ii) f is not differentiable for some $x \in (a,b)$ and
(b) how Rolle's Theorem **will succeed** even if f is not differentiable at $x = a$.
(This illustrates that f doesn't have to be differentiable at $x = a$ for Rolle's Theorem!)

Solution

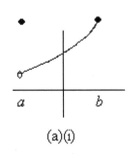
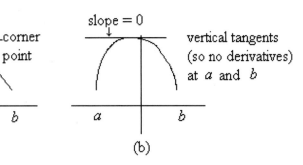

(a)(i) no zero slope   (a)(ii) no zero slope   (b) There is—**as there must be**—a zero slope!

The Mean Value Theorem

Suppose $y = f(x)$ is
(i) continuous on $[a,b]$ and
(ii) differentiable on (a,b).
Then there is **at least one** $c \in (a,b)$ such that
$$f'(c) = \frac{f(b) - f(a)}{b - a}.$$

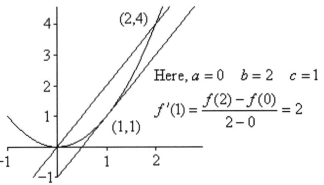

Here, $a = 0 \quad b = 2 \quad c = 1$
$$f'(1) = \frac{f(2) - f(0)}{2 - 0} = 2$$

Note: when $f(a) = f(b)$, the MVT becomes Rolle's Theorem.

Example 1) Verify The Mean Value Theorem with $f(x) = x^2 - 1$, $a = -1$, and $b = 2$.

Solution Since f is a polynomial, it is differentiable and continuous on $(-1, 2)$ and $[-1, 2]$, respectively. According to The Mean Value Theorem, we should be able to find $c \in (-1, 2)$ satisfying $f'(c) = \dfrac{f(2) - f(-1)}{2 - (-1)} = \dfrac{3 - 0}{3} = 1$.

Since $f'(x) = 2x$, solving $2c = 1$, we find $c = \dfrac{1}{2} \in (-1, 2)$.

Example 2) Sketch the graph of a function showing
(a) how the conclusion of The Mean Value Theorem **can fail** if
   (i) f is not continuous at a
   (ii) f is not differentiable for some $x \in (a,b)$ and
(b) how, if the hypothesis is satisfied, The Mean Value Theorem **will succeed** even if f is not differentiable at $x = a$. (This shows that $f'(a)$ is not necessary for The Mean Value Theorem to work!)

Solution

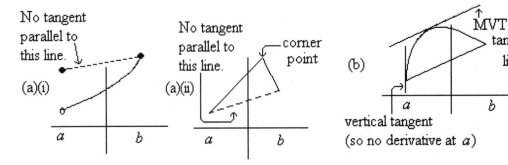

Two For You – Graph Sketching With Calculus: Vertical Tangent!

1) The function $y = x^{\frac{4}{3}} + 4x^{\frac{1}{3}}$ satisfies $y' = \frac{4}{3}\left(\dfrac{x+1}{x^{\frac{2}{3}}}\right)$ and $y'' = \frac{4}{9}\left(\dfrac{x-2}{x^{\frac{5}{3}}}\right)$. Using intercepts, intervals of increasing and decreasing, intervals of concave up and concave down, extreme points, and points of inflection, create a beautiful graph!

2) The function $y = (x+1)^{\frac{1}{3}}(x-1)^{\frac{2}{3}}$ satisfies $y' = \dfrac{x + \frac{1}{3}}{(x+1)^{\frac{2}{3}}(x-1)^{\frac{1}{3}}}$ and $y'' = \dfrac{-8}{9(x+1)^{\frac{5}{3}}(x-1)^{\frac{4}{3}}}$. Using intercepts, intervals of increasing and decreasing, intervals of concave up and concave down, extreme points, and points of inflection, create a beautiful graph!

Answers

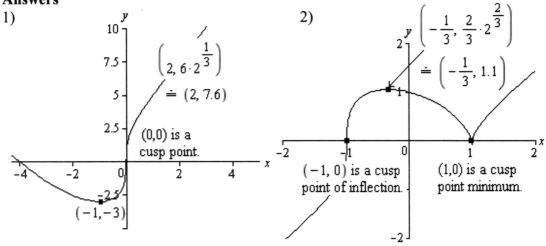

Two For You – Estimating Using the Differential

1) Estimate $\sqrt{3.98}$ using the differential.
(Hint: this is just like the example above but use $dx = -0.02$.)

2) Give the best choice for $f(x)$, x, and dx in order to estimate $\dfrac{1}{26^{1/3}}$ using the differential.

Answers 1) 1.995 2) $f(x) = \dfrac{1}{x^{1/3}}$, $x = 27$, $dx = -1$

Two For You – Rolle's Theorem

1) Verify Rolle's Theorem for the function $f(x) = x^3 - x + 3$, with $a = 0$ and $b = 1$.

2) Sketch a function that illustrates how the conclusion of Rolle's Theorem **can fail**, that is, there will be no 0 derivative, if f is not continuous for some $x \in (a,b)$.

Answers 1) $f'\left(\dfrac{1}{\sqrt{3}}\right) = 0$ and $\dfrac{1}{\sqrt{3}} \in (0,1)$, as required.

2)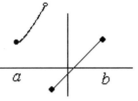

Two For You – The Mean Value Theorem

1) Verify The Mean Value Theorem with $f(x) = x^3 + 3$, $a = -1$, and $b = 2$.

2) Sketch a function **not continuous** at some $x \in (a,b)$ for which the conclusion of the MVT **fails**.

Answers 1) $f'(1) = 3$ and $1 \in (-1, 2)$, as required.   2)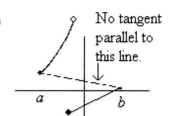

Derivatives: The Product Rule

Think of *First* and *Second*:

$$\frac{d}{dx}(FS) = FS' + SF' \quad \text{Note: the order is NOT important!}$$

Example 1) Find $\frac{dy}{dx}$ if $y = (x^3 + x)(\sin x - 4)$.

Solution $\frac{dy}{dx} = \underset{[F]}{(x^3+x)}\underset{[S']}{\cos x} + \underset{[S]}{(\sin x - 4)}\underset{[F']}{(3x^2+1)}$

$= x^3 \cos x + x \cos x + 3x^2 \sin x - 12x^2 + \sin x - 4$

Here is the product rule combined with the **chain rule** (reviewed in the next section!)

Example 2) Find $\frac{dy}{dx}$ if $y = x(e^x + \ln x)^5$.

Solution $\frac{dy}{dx} = x\left(5(e^x + \ln x)^4\left(e^x + \frac{1}{x}\right)\right) + (e^x + \ln x)^5 (1)$

$= (e^x + \ln x)^4 \left(5x\left(e^x + \frac{1}{x}\right) + (e^x + \ln x)\right)$

$= (e^x + \ln x)^4 \left(5xe^x + 5 + e^x + \ln x\right)$

Example 3) Find the formula for the derivative of the product of three functions, that is, find $\frac{dy}{dx}$ when $y = f(x)g(x)h(x)$.

Solution Group two of the functions together and use the basic product rule.
Writing $y = (f(x)g(x))h(x)$,

$\frac{dy}{dx} = (f(x)g(x))h'(x) + h(x)(f(x)g(x))'$

$= (f(x)g(x))h'(x) + h(x)(f(x)g'(x) + g(x)f'(x))$

$= f(x)g(x)h'(x) + f(x)h(x)g'(x) + g(x)h(x)f'(x)$

In short: $\boxed{(fgh)' = f'gh + fg'h + fgh'}$

Derivatives: The Chain Rule

Example 1) Find $\dfrac{dy}{dx}$ if $y = (\sin x + x)^3$.

Solution Think of this as $y = (inside)^3$.

The chain rule tells us to take the derivative of the **outside** and multiply by the derivative of the **inside**. Keep taking the derivative until you "get to the derivative with respect to x." So…

$$\frac{dy}{dx} = \frac{d(inside^3)}{d(inside)} \frac{d(inside)}{dx} = 3(inside)^2 \frac{d(inside)}{dx} = 3(\sin x + x)^2 (\cos x + 1)$$

Now, repeating myself in my (slowly approaching **late**) middle age, we keep going and going and going until we finally get down to x. If you can do this next example, you are a chain rule pro!

Example 2) Find $\dfrac{dy}{dx}$ if $y = \sqrt{x + \sqrt{x + \sqrt{x}}}$. Here we are going to have go inside **once! twice!! three!!! times!!!!**

Solution I find it much easier to handle an example like this if I rewrite it as

$$y = \sqrt{x + \sqrt{x + \sqrt{x}}} = \left(x + \left(x + x^{1/2}\right)^{1/2}\right)^{1/2}.$$

We are going to take the derivative of the outside and then multiply by the derivative of the **FIRST** inside and then go inside again and, for the final $x^{1/2}$, a third time!

$$\frac{dy}{dx} = \frac{1}{2}\left(x + \left(x + x^{1/2}\right)^{1/2}\right)^{-1/2} \left(1 + \frac{1}{2}\left(x + x^{1/2}\right)^{-1/2} \left(1 + \frac{1}{2}x^{-1/2}\right)\right)$$

$$= \frac{1}{2\sqrt{x + \sqrt{x + \sqrt{x}}}}\left(1 + \frac{1}{2\sqrt{x + \sqrt{x}}}\left(1 + \frac{1}{2\sqrt{x}}\right)\right)$$

Derivatives: The Quotient Rule

Think of **Top** (numerator) and **Bottom** (denominator):

$$\boxed{\frac{d}{dx}\left(\frac{T}{B}\right) = \frac{BT' - TB'}{B^2}}$$

Note the "−". **Note** the order. **Note**, no $B' = \frac{dB}{dx}$ in the bottom!

Example 1) Find $\frac{dy}{dx}$ if $y = \frac{4x^2}{x^2 - \sin x} = 4\left(\frac{x^2}{x^2 - \sin x}\right)$.

Solution $\frac{dy}{dx} \stackrel{\text{Keep the constant 4 outside!}}{=} 4\left(\frac{\overset{B}{(x^2 - \sin x)}\overset{T'}{(2x)} - \overset{T}{x^2}\overset{B'}{(2x - \cos x)}}{\underset{B^2}{(x^2 - \sin x)^2}}\right)$

$= 4\left(\frac{2x^3 - 2x\sin x - 2x^3 + x^2\cos x}{(x^2 - \sin x)^2}\right) = \frac{4(x^2\cos x - 2x\sin x)}{(x^2 - \sin x)^2}$

Here is a common use of the chain and quotient rules together:

Example 2) Find $\frac{dy}{dx}$ if $y = \left(\frac{3x+5}{4x+7}\right)^5$.

Solution $\frac{dy}{dx} = 5\left(\frac{3x+5}{4x+7}\right)^4 \left(\frac{(4x+7)(3) - (3x+5)(4)}{(4x+7)^2}\right)$

$\stackrel{\text{Combine the }(4x+7)\text{ factors in the denominator.}}{=} 5\left(\frac{(3x+5)^4}{(4x+7)^6}\right)(12x + 21 - 12x - 20) = \frac{5(3x+5)^4}{(4x+7)^6}$

Alternate Solution Use the product rule with the chain rule:

$y = \left(\frac{3x+5}{4x+7}\right)^5 = (3x+5)^5(4x+7)^{-5}$

$\frac{dy}{dx} = (3x+5)^5(-5)(4x+7)^{-6}(4) + (4x+7)^{-5}(5)(3x+5)^4(3)$

$\stackrel{\text{Take out the common factors.}}{=} 5(3x+5)^4(4x+7)^{-6}(-4(3x+5) + 3(4x+7))$ *

$= 5(3x+5)^4(4x+7)^{-6}(-12x - 20 + 12x + 21) = \frac{5(3x+5)^4}{(4x+7)^6}$

*Note that 4 was the lower of the two exponents on $(3x+5)$ and -6 was the lower exponent of the two exponents on $(4x+7)$.

Derivatives: Implicit Differentiation

Don't panic! Implicit Differentiation is just an application of the **chain rule** in disguise.

Example 1) Let $xy + \sin y = 4$. (a) Find $\dfrac{dy}{dx}$. (b) Find x and $\dfrac{dy}{dx}$ when $y = \pi$.

Solution It is **impossible** to rewrite this equation with y on the left side and only terms involving x on the right side. (Try to isolate y. Futile!) Instead, to find $\dfrac{dy}{dx}$, we take the derivative with respect to x directly from the equation using the **GOLDEN RULE** of math equations:

> What you do to one side you do to the other! *

Note that we need the **product rule** to deal with the xy term.

$$\frac{d(Left\ Side)}{dx} = \frac{d(Right\ Side)}{dx} \quad \therefore\ x\frac{dy}{dx} + y(1) + \cos y \frac{dy}{dx} = 0.$$

Now factor out the $\dfrac{dy}{dx}$ terms and solve: $\dfrac{dy}{dx}(x + \cos y) = -y \Rightarrow \dfrac{dy}{dx} = \dfrac{-y}{x + \cos y}$

(b) $y = \pi \Rightarrow \pi x + 0 = 4 \Rightarrow x = \dfrac{4}{\pi}$.

At the point $\left(\dfrac{4}{\pi}, \pi\right)$, $\dfrac{dy}{dx} = \dfrac{-\pi}{\dfrac{4}{\pi} + \cos(\pi)} = \dfrac{-\pi}{\dfrac{4}{\pi} - 1} \overset{\text{Multiply the top and the bottom by }\pi.}{=} \dfrac{-\pi^2}{4 - \pi}$

Example 2) Find $\dfrac{dy}{dx}$ if $x^2 y^3 + 2x + 3y = \sin(xy) + 4$.

Solution $x^2(3y^2)\dfrac{dy}{dx} + 2xy^3 + 2 + 3\dfrac{dy}{dx} = \cos(xy)\left(x\dfrac{dy}{dx} + y\right)$

> Now bring $\dfrac{dy}{dx}$ terms to the left and factor out $\dfrac{dy}{dx}$.
> All other terms go to, or stay on, the right side.

$\dfrac{dy}{dx}\left(3x^2 y^2 + 3 - x\cos(xy)\right) = y\cos(xy) - 2 - 2xy^3$

$\therefore\ \dfrac{dy}{dx} = \dfrac{y\cos(xy) - 2 - 2xy^3}{3x^2 y^2 + 3 - x\cos(xy)}$

*Would you consider this a precise math analog of the usual **GOLDEN RULE**?

Two For You – Derivatives: The Product Rule

1) Find $\dfrac{dy}{dx}$ for each of the following:

(a) $y = x^4(x^5 + 10)^6$ (b) $y = (5x + 4)(\sin x)(\ln x)$

2) Write the formula for $\dfrac{dy}{dx}$ if $y = f_1 f_2 f_3 f_4$.

Answers 1)(a) $2x^3(x^5 + 10)^5(17x^5 + 20)$

(b) $\dfrac{(5x+4)(\sin x)}{x} + (5x+4)(\ln x)(\cos x) + 5(\sin x)(\ln x)$

2) $f_1' f_2 f_3 f_4 + f_1 f_2' f_3 f_4 + f_1 f_2 f_3' f_4 + f_1 f_2 f_3 f_4'$

Two For You – Derivatives: The Chain Rule

Find $\dfrac{dy}{dx}$ for each of the following:

1) $y = \sin(x^3 + 3x^2 + 1)$

2) $y = \dfrac{1}{(2x-1)^{3/2}}$ (Hint: first rewrite this as $y = (2x-1)^{-3/2}$.)

Answers 1) $(3x^2 + 6x)\cos(x^3 + 3x^2 + 1)$ 2) $-3(2x-1)^{-5/2}$ $\overset{\text{optional back of book answer}}{=} \dfrac{-3}{(2x-1)^{5/2}}$

Two For You – Derivatives: The Quotient Rule

Find $\dfrac{dy}{dx}$ for each of the following:

1) $y = \dfrac{x^2+1}{e^x+x^2}$ 2) $y = \left(\dfrac{x^4+1}{x^4-1}\right)^2$

Answers 1) $\dfrac{2xe^x - x^2e^x - e^x - 2x}{(e^x+x^2)^2}$ 2) $\dfrac{-16x^3(x^4+1)}{(x^4-1)^3}$

Two For You – Derivatives: Implicit Differentiation

1)(a) Find $\dfrac{dy}{dx}$ if $e^{x+y} + 5x + 2y = 1$. (b) Find $\dfrac{dy}{dx}$ at $(0,0)$.

2) Find $\dfrac{dy}{dx}$ if $y \sin x + y = x + \tan y$.

Answers 1)(a) $\dfrac{-(e^{x+y}+5)}{e^{x+y}+2}$ (b) -2 2) $\dfrac{1 - y\cos x}{\sin x + 1 - \sec^2 y}$

Derivatives: Implicit Differentiation Second Derivative

Here the key is one simple fact: $\dfrac{d}{dx}\left(\dfrac{dy}{dx}\right)=\dfrac{d^2y}{dx^2}$, that is, the derivative with respect to x of the first derivative (with respect to x) is the second derivative (with respect to x).

Also, the second derivative may seem to get mechanically scary. It is not implicit differentiation that causes this, **but GRADE 5 fractions!**

Example 1) Find $\dfrac{dy}{dx}$ and $\dfrac{d^2y}{dx^2}$ if $e^y - xy = 1$.

Solution Find the first derivative implicitly.

$$e^y \frac{dy}{dx} - x\frac{dy}{dx} - y(1) = 0 \Rightarrow \frac{dy}{dx}(e^y - x) = y \text{ and so } \frac{dy}{dx} = \frac{y}{e^y - x}$$

Now, we use the quotient rule to find $\dfrac{d^2y}{dx^2}$.

$$\frac{d^2y}{dx^2} = \frac{(e^y - x)\dfrac{dy}{dx} - y(e^y \dfrac{dy}{dx} - 1)}{(e^y - x)^2} \quad \boxed{\text{Factor out } \tfrac{dy}{dx} \text{ and expand the numerator.}} = \frac{\dfrac{dy}{dx}(e^y - x - ye^y) + y}{(e^y - x)^2}$$

$$\boxed{\text{We already know } \tfrac{dy}{dx}=\tfrac{y}{e^y-x}!} = \frac{\left(\dfrac{y}{e^y - x}\right)(e^y - x - ye^y) + y}{(e^y - x)^2}$$

$$\boxed{\text{Get a common denominator in the numerator.}} = \frac{y(e^y - x - ye^y) + y(e^y - x)}{(e^y - x)} \cdot \frac{1}{(e^y - x)^2}$$

$$\boxed{\text{Expand the numerator...}} = \frac{ye^y - xy - y^2e^y + ye^y - xy}{(e^y - x)^3} \quad \boxed{\text{...and collect like terms.}} = \frac{2ye^y - 2xy - y^2e^y}{(e^y - x)^3}$$

Alternate Solution Differentiate $\dfrac{dy}{dx}(e^y - x) = y$ using the Product Rule.

$$\frac{dy}{dx}(e^y \frac{dy}{dx} - 1) + (e^y - x)\frac{d^2y}{dx^2} = \frac{dy}{dx} \text{ and so } (e^y - x)\frac{d^2y}{dx^2} = 2\frac{dy}{dx} - e^y\left(\frac{dy}{dx}\right)^2$$

$$= 2\frac{y}{e^y - x} - e^y\left(\frac{y}{e^y - x}\right)^2 \quad \boxed{\text{Common Denominator!}} = \frac{2ye^y - 2xy - y^2e^y}{(e^y - x)^2}$$

and so, $\dfrac{d^2y}{dx^2} = \dfrac{2ye^y - 2xy - y^2e^y}{(e^y - x)^3}$

Easy Integrals/Anti-Derivatives

Okay, this page is the most basic of the basic—the anti-derivative/integral formulas you learn in your first calculus semester applied with **NO TRICKS, NO TWISTS, NO COMPLICATIONS**. The diabolical stuff, and the means to deal with it, comes later!

$$\int 0\,dx = C \qquad \int 1\,dx = x + C \qquad \int m\,dx = mx + C$$

$$\int f(x) \pm g(x)\,dx = \int f(x)\,dx \pm \int g(x)\,dx \qquad \int cf(x)\,dx = c\int f(x)\,dx$$

$$\int x^n\,dx = \frac{x^{n+1}}{n+1} + C,\ n \neq -1 \qquad \int \frac{1}{x}\,dx = \int x^{-1}\,dx = \ln|x| + C$$

$$\int \sin x\,dx = -\cos x + C \qquad \int \cos x\,dx = \sin x + C$$

$$\int \tan x\,dx = -\ln|\cos x| + C \qquad \int \cot x\,dx = \ln|\sin x| + C$$

$$\int \csc x\,dx = \ln|\csc x - \cot x| + C \qquad \int \sec x\,dx = \ln|\sec x + \tan x| + C$$

$$\int e^x\,dx = e^x + C \qquad \int a^x\,dx = \frac{a^x}{\ln a} + C$$

Example 1) Evaluate the following integrals:

(a) $\int 1 + x^2 + x^{-5/3}\,dx$ (b) $\int \sin x + 2\csc x + \frac{1}{3}e^x + 10^x\,dx$

Solution

(a) $\int 1 + x^2 + x^{-5/3}\,dx \overset{\text{for } n\neq -1,\ \int x^n dx = \frac{x^{n+1}}{n+1}+C}{=} x + \frac{x^3}{3} + \frac{x^{-2/3}}{-\frac{2}{3}} + C = x + \frac{x^3}{3} - \frac{3x^{-2/3}}{2} + C$

(b) $\int \sin x + 2\csc x + \frac{1}{3}e^x + 10^x\,dx = -\cos x + 2\ln|\csc x - \cot x| + \frac{1}{3}e^x + \frac{10^x}{\ln 10} + C$

Example 2) If $\int f(x)\,dx = 3x^2 + C$ and $\int g(x)\,dx = 2\sin x - e^x + D$, find $\int 3f(x) - 2g(x)\,dx$.

Solution $\int 3f(x) - 2g(x)\,dx = 3\int f(x)\,dx - 2\int g(x)\,dx = 3(3x^2 + C) - 2(2\sin x - e^x + D)$

$= 9x^2 - 4\sin x + 2e^x + 3C - 2D \overset{\text{Replace the constant } 3C-2D \text{ with the single constant } E.}{=} 9x^2 - 4\sin x + 2e^x + E$

Easy Integrals that Need a Little Tweaking

Here is a variety of really easy integrals—except you have to do something first to see why they are so easy.

Example 1) Evaluate the following integrals:

(a) $\int (x^3 + 3)^2 \, dx$ (b) $\int (x^{1/2} + 1)(2x + 5) \, dx$ (c) $\int \dfrac{x^4 - 7}{x^2} \, dx$

(d) $\int \dfrac{1}{\sec x} \, dx$ (e) $\int \dfrac{1}{e^x} \, dx$

Solution (a) EXPAND: $\int (x^3 + 3)^2 \, dx = \int x^6 + 6x^3 + 9 \, dx = \dfrac{x^7}{7} + \dfrac{3x^4}{2} + 9x + C$

Note how unpleasant this example would be if changed to $\int (x^3 + 3)^{20} \, dx$. Note how much MORE unpleasant it would be if changed to $\int (x^3 + 3)^{2/3} \, dx$! The first you could expand, but would you want to? The second, you **can't** expand, at least not with your current math toolkit! **But test your understanding of the Chain Rule in Reverse (CRIR:** See page 192.) :

both $\int \boxed{x^2}(x^3 + 3)^{20} \, dx$ and $\int \boxed{x^2}(x^3 + 3)^{2/3} \, dx$ are **easy**!

$\int x^2 (x^3 + 3)^{20} \, dx \overset{\text{Adjust by 3.}}{=} \dfrac{1}{3} \int 3x^2 (x^3 + 3)^{20} \, dx \overset{\text{CRIR!}}{=} \dfrac{1}{3}\left(\dfrac{(x^3 + 3)^{21}}{21}\right) + C = \dfrac{(x^3 + 3)^{21}}{63} + C$

$\int x^2 (x^3 + 3)^{2/3} \, dx \overset{\text{Adjust by 3.}}{=} \dfrac{1}{3} \int 3x^2 (x^3 + 3)^{2/3} \, dx \overset{\text{CRIR!}}{=} \dfrac{1}{3}\left(\dfrac{(x^3 + 3)^{5/3}}{5/3}\right) + C = \dfrac{(x^3 + 3)^{5/3}}{5} + C$

(b) $\int (x^{1/2} + 1)(2x + 5) \, dx \overset{\text{Expand!}}{=} \int 2x^{3/2} + 5x^{1/2} + 2x + 5 \, dx = \dfrac{4}{5}x^{5/2} + \dfrac{10}{3}x^{3/2} + x^2 + 5x + C$

(c) $\int \dfrac{x^4 - 7}{x^2} \, dx \overset{\text{Make separate fractions!}}{=} \int x^2 - 7x^{-2} \, dx = \dfrac{x^3}{3} - \dfrac{7x^{-1}}{-1} + C = \dfrac{x^3}{3} + \dfrac{7}{x} + C$

(d) $\int \dfrac{1}{\sec x} \, dx \overset{\text{This is just a case of the teacher being sneaky!}}{=} \int \cos x \, dx = \sin x + C$

(e) $\int \dfrac{1}{e^x} \, dx \overset{\text{This is just a case of the teacher being sneaky again!}}{=} \int e^{-x} \, dx \overset{\text{CRIR: Adjust by }-1.}{=} -\int -e^{-x} \, dx = -e^{-x} + C$

The Chain Rule in Reverse: No Adjustments Needed

Many students find this **very** hard but in fact it is pretty easy. **The chain rule in reverse:**

BASIC	CHAIN RULE IN REVERSE
$\int f'(x)\,dx = f(x)+C$	$\int f'(u)\dfrac{du}{dx}\,dx = f(u)+C$

It may be gross but I tell my students that when taking the derivative of $y = f(u)$...

> Please read $f(u)$ as "f **AT** u"; otherwise, we go to a whole other level of gross!

...with respect to x, the chain rule "**spits out**" du/dx. So, when taking the anti-derivative or integral with respect to x, the du/dx is "**sucked**" back inside the u!

Think **"DOUBLE S"**: **S**pit for the derivative, **S**uck for the integral. Here are all the basic formulas, modified to show the CRIR.

$$\int u^n \frac{du}{dx}\,dx \overset{[n \neq -1]}{=} \frac{u^{n+1}}{n+1}+C, \qquad \int \frac{1}{u}\frac{du}{dx}\,dx = \int u^{-1}\frac{du}{dx}\,dx = \ln|u|+C$$

$$\int \sin u \frac{du}{dx}\,dx = -\cos u + C \qquad \int \cos u \frac{du}{dx}\,dx = \sin u + C \qquad \int \tan u \frac{du}{dx}\,dx = -\ln|\cos u|+C$$

$$\int \cot u \frac{du}{dx}\,dx = \ln|\sin u|+C \qquad \int e^u \frac{du}{dx}\,dx = e^u + C \qquad \int a^u \frac{du}{dx}\,dx = \frac{a^u}{\ln a}+C$$

$$\int \csc u \frac{du}{dx}\,dx = \ln|\csc u - \cot u|+C \qquad \int \sec x \frac{du}{dx}\,dx = \ln|\sec u + \tan u|+C$$

Example 1) Evaluate, using the chain rule in reverse. Identify u and $\dfrac{du}{dx}$.

(a) $\int (x^3+1)^{10}(3x^2)\,dx$ (b) $\int e^{\tan x} \sec^2 x \,dx$ (c) $\int \cos(x^{1/2})\left(\dfrac{1}{2}x^{-1/2}\right)dx$

(d) $\int \dfrac{2x}{x^2+1}\,dx$

Solution (a) $\int (x^3+1)^{10}(3x^2)\,dx \quad \boxed{u=x^3+1,\ \tfrac{du}{dx}=3x^2} \quad = \dfrac{(x^3+1)^{11}}{11}+C$

(b) $\int e^{\tan x} \sec^2 x \,dx \quad \boxed{u=\tan x,\ \tfrac{du}{dx}=\sec^2 x} \quad = e^{\tan x}+C$

(c) $\int \cos(x^{1/2})\left(\dfrac{1}{2}x^{-1/2}\right)dx \quad \boxed{u=x^{1/2},\ \tfrac{du}{dx}=\tfrac{1}{2}x^{-1/2}} \quad = \sin(x^{1/2})+C$

(d) $\int \dfrac{2x}{x^2+1}\,dx \quad \boxed{u=x^2+1,\ \tfrac{du}{dx}=2x} \quad = \ln(x^2+1)+C$ (Note: $x^2+1>0$, so we don't need "$|\ |$".)

Two For You – Implicit Differentiation Second Derivative

Find $\dfrac{dy}{dx}$ and $\dfrac{d^2y}{dx^2}$ for each of the following: 1) $x^3 + y^3 = 1$ 2) $y^2 + 2xy = 10$

(Hint: in each question, at the **last step**, use the original equation!)

Answers 1) $\dfrac{dy}{dx} = -\dfrac{x^2}{y^2}$, $\dfrac{d^2y}{dx^2} = \dfrac{-2x}{y^5}$ 2) $\dfrac{dy}{dx} = -\dfrac{y}{x+y}$, $\dfrac{d^2y}{dx^2} = \dfrac{10}{(x+y)^3}$

Two For You – Easy Integrals/Anti-Derivatives

Evaluate the following integrals:

1) $\int -1 + 4x^9 - x^{-5/3}\, dx$

2) $\int \cot x - \sec x - 2^x \ln 2\, dx$ $\left(\text{Hint: } \int 2^x \ln 2\, dx = \ln 2 \left(\int 2^x\, dx\right)\right)$

Answers 1) $-x + \dfrac{2x^{10}}{5} + \dfrac{3x^{-2/3}}{2} + C$ 2) $\ln|\sin x| - \ln|\sec x + \tan x| - 2^x + C$

Two For You – Easy Integrals that Need a Little Tweaking

Evaluate the following integrals:

1) $\int (2x^2 - 1)^3 \, dx$ 2) $\int \dfrac{e^{3x} - 2e^x + e^{-x}}{2e^x} \, dx$

Answers 1) $\dfrac{8x^7}{7} - \dfrac{12x^5}{5} + 2x^3 - x + C$ 2) $\dfrac{1}{4}e^{2x} - x - \dfrac{1}{4e^{2x}} + C$

Two For You – The CRIR: No Adjustments Needed!

Evaluate using the **Chain Rule In Reverse**. For each, identify u and $\dfrac{du}{dx}$.

1) $\int (x^5 + 2x^2 + 1)^{2/3} (5x^4 + 4x) \, dx$ 2) $\int \sin(\sin x)(\cos x) \, dx$

Answers 1) $\dfrac{3}{5}(x^5 + 2x^2 + 1)^{5/3} + C$; $u = x^5 + 2x^2 + 1$, $\dfrac{du}{dx} = 5x^4 + 4x$

2) $-\cos(\sin x) + C$; $u = \sin x$, $\dfrac{du}{dx} = \cos x$

The Chain Rule in Reverse: Adjustments Needed BUT Don't Use Substitution!

Now let's use the **the chain rule in reverse** where we adjust by a "**multiplicative constant**". You can always pull a multiplicative constant outside the integral and you can adjust an integral by a multiplicative constant **providing you compensate**. The CRIR "sucks" in the *du/dx*. The key here is that you recognize **in advance** that you have both the *u* and the *du/dx*—at least up to the constant—in the integral. You know before you start that the CRIR will solve the problem.

Example 1) Evaluate, using the chain rule in reverse. In each, identify u and $\frac{du}{dx}$.

(a) $\int x^2(x^3+1)^{10}\, dx$ (b) $\int \frac{\cos\sqrt{x}}{\sqrt{x}}\, dx$ (c) $\int 3e^{\tan x}\sec^2 x\, dx$ (d) $\int \frac{x}{x^2+1}\, dx$

Solution

(a) $\int x^2(x^3+1)^{10}\, dx$ Here, $u = x^3+1$ and $\frac{du}{dx} = 3x^2$. We need to multiply by 3. To compensate, we divide by 3 **outside** the integral sign.

YOU CAN ALWAYS ADJUST BY A MULTIPLICATIVE CONSTANT.

$$\int x^2(x^3+1)^{10}\, dx = \frac{1}{3}\int 3x^2(x^3+1)^{10}\, dx = \frac{1}{3}\frac{(x^3+1)^{11}}{11} + C = \frac{1}{33}(x^3+1)^{11} + C$$

Question: Where has the $3x^2$ gone? Answer: **SUCKED INSIDE by the chain rule!**

(b) $\int \frac{\cos\sqrt{x}}{\sqrt{x}}\, dx$ Here, $u = \sqrt{x} = x^{1/2}$ and $\frac{du}{dx} = \frac{1}{2}x^{-1/2} = \frac{1}{2\sqrt{x}}$. We need to divide by 2. To compensate, we multiply by 2 outside the integral sign.

$$\int \frac{\cos\sqrt{x}}{\sqrt{x}}\, dx = 2\int \frac{1}{2}x^{-1/2}\cos(x^{1/2})\, dx = 2\sin(x^{1/2}) + C = 2\sin\sqrt{x} + C$$

(c) $\int 3e^{\tan x}\sec^2 x\, dx$ Here, $u = \tan x$ and $\frac{du}{dx} = \sec^2 x$, so no compensation needed.

But move the constant 3 outside: $\int 3e^{\tan x}\sec^2 x\, dx = 3\int e^{\tan x}\sec^2 x\, dx = 3e^{\tan x} + C$

(d) $\int \frac{x}{x^2+1}\, dx$ Here, $u = x^2+1$ and $\frac{du}{dx} = 2x$. Compensate with 2.

$$\int \frac{x}{x^2+1}\, dx = \frac{1}{2}\int \frac{2x}{x^2+1}\, dx \quad \boxed{x^2+1>0 \text{ so we don't need absolute value.}} = \frac{1}{2}\ln(x^2+1) + C$$

The Chain Rule in Reverse: Adjustments Needed and Using Substitution

Let's get something straight: I don't like using substitution for these problems. Why? Because EVERY TIME you adjust the constant to make things work just right as we did on the preceding topic, you consolidate further your understanding of the CRIR. It is so easy! Yet students find it so hard! The real key is recognizing that the questions are **"cooked"**! The du/dx term, up to a constant, **must** be present for most problems or the integral is, in **many** examples, too hard or even undo-able. Even so, this page is almost exactly the same as the preceding one, except here we will use substitution.

Example 1) Evaluate, using the chain rule in reverse. In each, identify u and du.

(a) $\int x^2 (x^3+1)^{10} \, dx$ (b) $\int \dfrac{\cos(\sqrt{x})}{\sqrt{x}} \, dx$ (c) $\int 3e^{\tan(x)} \sec^2(x) \, dx$ (d) $\int \dfrac{x}{x^2+1} \, dx$

Solution

(a) $\int x^2 (x^3+1)^{10} \, dx$ Here, $u = x^3 + 1$ and $\dfrac{du}{dx} = 3x^2$.

$\therefore du = 3x^2 \, dx$. We need to replace $x^2 \, dx$. Since $x^2 \, dx = \dfrac{1}{3} du$,

we have $\int x^2 (x^3+1)^{10} \, dx \;\underset{\text{Pull } \frac{1}{3} \text{ outside the integral.}}{=}\; \dfrac{1}{3} \int u^{10} \, du = \dfrac{1}{3} \dfrac{u^{11}}{11} + C \;\underset{u=x^3+1}{=}\; \dfrac{1}{33}(x^3+1)^{11} + C$

(b) $\int \dfrac{\cos(\sqrt{x})}{\sqrt{x}} \, dx$ Here, $u = \sqrt{x} = x^{1/2}$ and $\dfrac{du}{dx} = \dfrac{1}{2} x^{-1/2} = \dfrac{1}{2\sqrt{x}}$ $\therefore du = \dfrac{dx}{2\sqrt{x}}$

So $2\, du = \dfrac{dx}{\sqrt{x}}$ and $\int \dfrac{\cos(\sqrt{x})}{\sqrt{x}} \, dx = 2\int \cos(u) \, du = 2\sin(u) + C = 2\sin(\sqrt{x}) + C$

(c) $\int 3e^{\tan(x)} \sec^2(x) \, dx$ Here, $u = \tan(x)$ and   *Let's go right to the differential this time!* $du = \sec^2(x) \, dx$

and $\int 3e^{\tan(x)} \sec^2(x) \, dx = 3\int e^u \, du = 3e^u + C = 3e^{\tan(x)} + C$

(d) $\int \dfrac{x}{x^2+1} \, dx$ Here, $u = x^2 + 1$ and $du = 2x \, dx$ $\therefore \dfrac{du}{2} = x \, dx$ and

$\int \dfrac{x}{x^2+1} \, dx = \dfrac{1}{2}\int \dfrac{du}{u} = \dfrac{1}{2} \ln|u| + C \;\underset{x^2+1>0 \text{ so we don't need absolute value.}}{=}\; \dfrac{1}{2} \ln(x^2+1) + C

Substitution when the CRIR Doesn't Apply

Here is a question just waiting for the Chain Rule in Reverse: $\int x^3(x^4+1)^{1/2}\,dx$.
A simple adjustment of the multiplicative constant gives a final answer in two steps. Substitution not needed!

$$\int x^3(x^4+1)^{1/2}\,dx = \frac{1}{4}\int 4x^3(x^4+1)^{1/2}\,dx = \frac{1}{6}(x^4+1)^{3/2} + C$$

Now here is a question where the CRIR just doesn't apply: $\int x^2(x+3)^{1/2}\,dx$.

We do not have $\dfrac{du}{dx}$ up to a multiplicative constant. Also, we can't just expand because of the exponent "1/2". So, we make the $x+3$ the variable by substitution.

Example 1) Evaluate $\int x^2(x+3)^{1/2}\,dx$.

Solution Let $u = x+3$. Then $du = dx$ and $x = u - 3$. Therefore,

$$\int x^2(x+3)^{1/2}\,dx = \int (u-3)^2 u^{1/2}\,du \stackrel{\text{Now expand.}}{=} \int (u^2 - 6u + 9)u^{1/2}\,du = \int u^{5/2} - 6u^{3/2} + 9u^{1/2}\,du$$

$$= \frac{2}{7}u^{7/2} - \frac{12}{5}u^{5/2} + 6u^{3/2} + C \stackrel{\text{Resubstitute for } u \text{ in terms of } x.}{=} \frac{2}{7}(x+3)^{7/2} - \frac{12}{5}(x+3)^{5/2} + 6(x+3)^{3/2} + C$$

Example 2) Evaluate $\int \dfrac{x+3}{2x-1}\,dx$.

Solution Let $u = 2x - 1$. Then $du = 2dx$ and so $dx = \dfrac{du}{2}$.

Also, $2x = u + 1$ and so $x = \dfrac{u+1}{2}$. Therefore,

$$\int \frac{\frac{u+1}{2} + 3}{u}\,\frac{du}{2} \stackrel{\text{Get a common denominator in the top.}}{=} \int \frac{\left(\frac{u+1+6}{2}\right)}{2u}\,du \stackrel{\text{Simplify the fraction. Remember: } \frac{\left(\frac{a}{b}\right)}{c} = \frac{a}{b} \times \frac{1}{c} = \frac{a}{bc}}{=} \int \frac{u+7}{4u}\,du \stackrel{\text{I like to pull the constant outside the integral.}}{=} \frac{1}{4}\int \frac{u+7}{u}\,du$$

$$\stackrel{\text{Make separate fractions.}}{=} \frac{1}{4}\int 1 + \frac{7}{u}\,du = \frac{1}{4}(u + 7\ln|u|) + C \stackrel{\text{Substitute } u = 2x-1.}{=} \frac{1}{4}(2x - 1 + 7\ln|2x-1|) + C$$

$$\stackrel{\text{optional}}{=} \frac{x}{2} - \frac{1}{4} + \frac{7}{4}\ln|2x-1| + C \stackrel{\text{optional: } D = C - \frac{1}{4}}{=} \frac{x}{2} + \frac{7}{4}\ln|2x-1| + D$$

The Chain Rule in Reverse: Products of Trig Funcions

This section is all about the Chain Rule in Reverse, $\int f'(u)\dfrac{du}{dx}dx = f(u)+ C$, specifically applied to trig products. $f(u)$ will be a trig function and $f'(u)$ will be present, possibly in hiding.

Example 1) $\int \sin^m(x)\cos^n(x)\,dx$, where at least one of m and n is an odd positive integer.

Evaluate, using the chain rule in reverse: (a) $\int \cos^3(x)\,dx$ (b) $\int \sin^5(x)\cos^2(x)\,dx$

Solution (a) $\int \cos^3(x)\,dx = \int \cos^2(x)\cos(x)\,dx = \int (1-\sin^2(x))\cos(x)\,dx$

$\boxed{\int (1-u^2)\dfrac{du}{dx}dx = \ldots}$
$= \int \cos(x) - \sin^2(x)\cos(x)\,dx$ $\boxed{\ldots u - \dfrac{u^3}{3} + C}$ $= \sin(x) - \dfrac{\sin^3(x)}{3} + C$

(b) $\int \sin^5(x)\cos^2(x)\,dx = \int \sin^4(x)\cos^2(x)\sin(x)\,dx = \int \left((1-\cos^2(x))\right)^2 \cos^2(x)\sin(x)\,dx$

$\boxed{\text{Expand } ((1-\cos^2(x))^2 \cos^2(x) \text{ to get powers of cos. Change } \sin(x) \text{ to } -\sin(x) \text{ and compensate.}}$

$= -\int \left(\cos^2(x) - 2\cos^4(x) + \cos^6(x)\right)(-\sin(x))\,dx$

$= -\dfrac{\cos^3(x)}{3} + \dfrac{2\cos^5(x)}{5} - \dfrac{\cos^7(x)}{7} + C$

Example 2) $\int \tan^m(x)\sec^n(x)\,dx$, where m is an odd $\in \mathbb{Z}^+$ or n is even $\in \mathbb{Z}^+$.

(Note: This method also works for $\int \cot^m(x)\csc^n(x)\,dx$.)

Evaluate, using the chain rule in reverse: (a) $\int \tan^3(x)\sec(x)\,dx$ (b) $\int \tan^4(x)\sec^4(x)\,dx$

Solution (a) $\int \tan^3(x)\sec(x)\,dx = \int \tan^2(x)\tan(x)\sec(x)\,dx = \int (\sec^2(x)-1)\tan(x)\sec(x)\,dx$

$\boxed{\int (u^2-1)\left(\dfrac{du}{dx}\right)dx = \dfrac{u^3}{3} - u + C}$
$= \dfrac{\sec^3(x)}{3} - \sec(x) + C$

(b) $\int \tan^4(x)\sec^4(x)\,dx = \int \tan^4(x)\sec^2(x)\sec^2(x)\,dx = \int \tan^4(x)\left(1+\tan^2(x)\right)\sec^2(x)\,dx$

$\boxed{\text{Expand } \tan^2(x)(1+\tan^2(x)) \text{ to get powers of tan.}}$

$= \int \left(\tan^4(x) + \tan^6(x)\right)(\sec^2(x))\,dx = \dfrac{\tan^5(x)}{5} + \dfrac{\tan^7(x)}{7} + C$

Other trig products to watch for:

$\int \sin^2(x)\,dx$ or $\int \cos^2(x)\,dx$: Use $\sin^2(x) = \dfrac{1-\cos(2x)}{2}$ and $\cos^2(x) = \dfrac{1+\cos(2x)}{2}$.

$\int \sec^3(x)\,dx$: Use integration by parts. (See page 203.)

Two For You – The CRIR: Adjustments Needed

Evaluate: 1) $\int (x^4+1)\sin(x^5+5x)\,dx$ 2) $\int e^{\sec(2x+1)}\sec(2x+1)\tan(2x+1)\,dx$

Answers 1) $-\dfrac{1}{5}\cos(x^5+5x)+C$ 2) $\dfrac{1}{2}e^{\sec(2x+1)}+C$

Three For You – The CRIR: Adjustments Needed & Substitution

Evaluate:

1) $\int \dfrac{x^3-\csc^2 x}{x^4+4\cot x}\,dx$ 2) $\int e^{e^x+x}\,dx$ (Hint: $e^{e^x+x}=e^{e^x}e^x$)   3) $\int \dfrac{x}{(5x^2+1)^4}\,dx$

Answers

1) $\dfrac{1}{4}\ln|x^4+4\cot x|+C$ 2) $e^{e^x}+C$ 3) $-\dfrac{1}{30(5x^2+1)^3}+C$

Two For You - Substitution when the CRIR Doesn't Apply

Evaluate the following integrals: 1) $\int (3x-1)(x+5)^{1/3}\,dx$ 2) $\int \dfrac{2x+x^2}{x+1}\,dx$

Answers 1) $\dfrac{9}{7}(x+5)^{7/3} - 12(x+5)^{4/3} + C$ 2) $\dfrac{x^2}{2} + x - \ln|x+1| + C$

Three For You – The CRIR: Products of Trig Functions

Evaluate:

1) $\int \sin^5(x)\,dx$ 2) $\int \sec^3(x)\tan^5(x)\,dx$

3) $\int \csc^3(x)\cot^5(x)\,dx$ (Hint: use $\cot^2(x) = \csc^2(x) - 1$ and $\dfrac{d(\csc(x))}{dx} = -\csc(x)\cot(x)$.)

Answers

1) $-\cos(x) + \dfrac{2}{3}\cos^3(x) - \dfrac{\cos^5(x)}{5} + C$ 2) $\dfrac{\sec^7(x)}{7} - \dfrac{2\sec^5(x)}{5} + \dfrac{\sec^3(x)}{3} + C$

3) $-\dfrac{\csc^7(x)}{7} + \dfrac{2\csc^5(x)}{5} - \dfrac{\csc^3(x)}{3} + C$

Integration by Parts (I by P): $\int u\,dv = uv - \int v\,du$

This is the first "**serious** integration technique" we study in calculus. The key is the ...

> **...Integration by Parts Strategy**
> Choose u so that du is an "easier" math expression than u. Choose dv so that (i) v is easy to find and (ii) v is no more difficult to work with than dv.

Example 1) Evaluate: $\int x e^x\,dx$

Solution Let $I = \int x e^x\,dx = \int u\,dv$ Choose u and dv obeying the strategy!

$$\text{Let } u = x \qquad dv = e^x\,dx$$

$$\therefore \underbrace{du = dx}_{du \text{ is easier to work with than } u.} \qquad \underbrace{v = e^x}_{v \text{ is easy to find and is no more difficult to work with than } dv.} \quad \text{(Don't add } C \text{ yet!)}$$

$$I \;\underset{\substack{I=uv-\int v\,du \\ uv=xe^x,\ \int v\,du=\int e^x\,dx}}{=}\; xe^x - \int e^x\,dx \;\underset{\text{Now add }C.}{=}\; xe^x - e^x + C$$

Example 2) Evaluate: $\int x^2 \cos(x)\,dx$

Solution This time we will need Integration by Parts **TWICE**!

Let $I = \int x^2 \cos(x)\,dx = \int u\,dv$. Choose u and dv!

$$\text{Let } u = x^2 \qquad dv = \cos(x)\,dx$$

$$\therefore \underbrace{du = 2x\,dx}_{du \text{ is easier to work with than } u!} \qquad \underbrace{v = \sin(x)}_{v \text{ is easy to find and is no more difficult to work with than } dv.} \quad \text{(Don't add } C \text{ yet!)}$$

$$I = x^2 \sin(x) - \underset{\substack{\text{Keep the constant} \\ \textbf{outside} \\ \text{the integral sign.}}}{2} \underset{\substack{\text{We will use "I by P" again} \\ \text{for this term.}}}{\int x \sin(x)\,dx}$$

$$\text{Let } u = x \quad dv = \sin(x)\,dx$$

(As we apply I by P again, we are "recycling" the letters u and dv.)

$$\therefore du = dx \quad v = -\cos(x)$$

$$\therefore I = x^2 \sin(x) - \underset{\substack{\text{We kept this consant outside} \\ \text{the bracket to avoid clutter.}}}{2} \underset{\text{This whole bracket replaces } \int x\sin(x)\,dx.}{\left(-x\cos(x) - \int -\cos(x)\,dx \right)}$$

$$= x^2 \sin(x) + 2x\cos(x) - 2\int \cos(x)\,dx$$

$$\underset{\text{Now add }C!}{=}\; x^2 \sin(x) + 2x\cos(x) - 2\sin(x) + C$$

"Circular" Integration by Parts

> **Integration by Parts CIRCULAR Strategy**
> Choose u so that du is "no harder" (not necessarily easier!) to work with than u. Choose dv so that (i) v is easy to find and (ii) is no more difficult to work with than dv. Cross your fingers.

Example 1) Evaluate: $\int e^x \cos(x)\,dx$

Solution Let $I = \int e^x \cos(x)\,dx = \int u\,dv$ Choose u and dv using the modified strategy!

$$\text{Let } u = e^x \qquad dv = \cos(x)\,dx$$
$$\therefore \quad du = e^x\,dx \qquad v = \sin(x) \quad \text{(Don't add } C \text{ yet!)}$$

[du is no harder to work with than u.]
[v is easy to find and is no more difficult to work with than dv.]

$$I = uv - \int v\,du$$
$$uv = e^x \sin(x),\ \int v\,du = \int e^x \sin(x)\,dx$$

$$I = e^x \sin(x) - \int e^x \sin(x)\,dx$$

[We will do "I by P" again for this term!]

Since we used $u = e^x$ the first time, we use $u = e^x$ again. See Example 2* for the reason letting $u = \sin(x)$ is a BAD IDEA!

$$*\text{Let } u = e^x \qquad dv = \sin(x)\,dx *$$
$$\therefore\ du = e^x\,dx \qquad v = -\cos(x)$$

[This whole bracket replaces $\int e^x \sin(x)\,dx$.]
[This is I, the original integral, reappearing. We have come full CIRCLE! CIRCULAR I BY P!]

$$\therefore I = e^x \sin(x) - \left(-e^x \cos(x) - \int -e^x \cos(x)\,dx\right) = e^x \sin(x) + e^x \cos(x) - \int e^x \cos(x)\,dx$$

So we now have $I = e^x \sin(x) + e^x \cos(x) - I$. **Solve** for I!

$$2I = e^x \sin(x) + e^x \cos(x)$$

[Solve for I. Add C now!]

$$I = \frac{1}{2}\left(e^x \sin(x) + e^x \cos(x)\right) + C = \frac{e^x}{2}\left(\sin(x) + \cos(x)\right) + C$$

Example 2) What goes wrong if we let $u = \sin(x)$ when we do "I by P" again at step *?

Solution Let's try it! Let $u = \sin(x) \qquad dv = e^x\,dx$

[This is a bad idea!]

$$\therefore\ du = \cos(x)\,dx \qquad v = e^x$$

[This whole bracket replaces $\int e^x \sin(x)\,dx$.]

$$\therefore I = e^x \sin(x) - \left(e^x \sin(x) - \int e^x \cos(x)\,dx\right) = e^x \sin(x) - e^x \sin(x) + I = I$$

We now have $I = I$! The **good news**: this is TRUE! The **bad news**: we haven't integrated. This method is like tying your shoelaces and then untying them. You may not have done anything wrong but you are not ready to walk!

Integration by Parts (I by P): The Tan—Sec Connection

Note: These are **very** popular I by P test questions! First, review page 202.

Special Toolbox for $\int \sec^3(x)\,dx$ **and** $\int \sec^5(x)\,dx$

$\tan^2(x) = \sec^2(x) - 1$ $\qquad u = \sec(x) \Rightarrow du = \sec(x)\tan(x)\,dx;$

$\int \sec^2(x)\,dx = \tan(x) + C$ $\qquad \int \sec(x)\,dx = \ln|\sec(x) + \tan(x)| + C$

The Chain Rule. $\dfrac{d(u^3)}{dx} = 3u^2 \dfrac{du}{dx}$

$u = \sec^3(x) = (\sec(x))^3 \Rightarrow du = 3(\sec(x))^2 \sec(x)\tan(x)\,dx = 3\sec^3(x)\tan(x)\,dx$

Example 1) Evaluate: $\int \sec^3(x)\,dx$

Solution Let $I = \int \sec^3(x)\,dx = \int u\,dv$ Choose u and dv using the I by P strategy!

$\qquad$ Let $u = \sec(x)$ $\qquad\qquad dv = \sec^2(x)\,dx$

$\qquad \therefore \quad du = \sec(x)\tan(x)\,dx \qquad\quad v = \tan(x)$ $\quad$ (Don't add C yet!)

[du is maybe a little harder to work with than u.] [v is easy to find and seems a little easier to work with than dv.]

$I = uv - \int v\,du$, $uv = \sec(x)\tan(x)$, $\int v\,du = \int \sec(x)\tan^2(x)\,dx$

[Now we will substitute $\tan^2(x) = \sec^2(x) - 1$.]

$I \quad = \quad \sec(x)\tan(x) - \int \sec(x)\,\tan^2(x)\,dx$

$\quad = \sec(x)\tan(x) - \int \sec(x)(\sec^2(x) - 1)\,dx$

[$\int f(x) - g(x)\,dx = \int f(x)\,dx - \int g(x)\,dx$]

$\quad = \sec(x)\tan(x) - \int \sec^3(x) - \sec(x)\,dx$

[This is I! Solve for I.] [This is in our Toolbox.]

$\quad = \sec(x)\tan(x) - \int \sec^3(x)\,dx + \int \sec(x)\,dx$

$\therefore 2I = \sec(x)\tan(x) + \ln|\sec(x) + \tan(x)|$

$\therefore I = \dfrac{1}{2}\left(\sec(x)\tan(x) + \ln|\sec(x) + \tan(x)|\right) + C$

Example 2) Show the first Integration by Parts steps in evaluating $\int \sec^5(x)\,dx$.

[We could have used $dv = \sec^3(x)\,dx$ and used the answer from Example 1, but v would be **very** complicated! Try it and I know I'll see you back here in ten minutes!]

Solution Let $u = \sec^3(x)$ $\qquad\qquad\qquad\qquad dv = \sec^2(x)\,dx$

[from the Toolbox]

$\therefore du = 3\sec^3(x)\tan(x)\,dx \qquad\qquad v = \tan(x)$ (Don't add C yet!)

[Keep the constant outside to avoid "integral" clutter.] [Now, as in Example 1, we will substitute $\tan^2(x) = \sec^2(x) - 1$.]

$I = \sec^3(x)\tan(x) - 3\int \sec^3(x)\,\tan^2(x)\,dx$

Now continue as in Example 1. This time, you will arrive at $4I = \ldots$(See **Two for you.**)

Integration by Trigonometric Substitution: Sin

We use the sine substitution with the expression $a^2 - b^2x^2$, with a, b both positive. We set $bx = a\sin(\theta)$. Then $a^2 - b^2x^2 = a^2 - a^2\sin^2(\theta) = a^2(1 - \sin^2(\theta)) = a^2\cos^2(\theta)$. With a little integration luck, a complicated algebraic integral will become a simple trig integral. For simplicity, assume $0 < \theta < \dfrac{\pi}{2}$, ie., θ is in the first quadrant.

Example 1) Evaluate $\int \sqrt{1-x^2}\, dx$.

Solution Here, $a = 1$ and $b = 1$. Let $x = \sin(\theta)$. Then $dx = \cos(\theta)\, d\theta$.

$$\sqrt{1-x^2} = \sqrt{1-\sin^2(\theta)} = \sqrt{\cos^2(\theta)} = |\cos(\theta)| \overset{\boxed{0<\theta<\frac{\pi}{2}}}{=} \cos(\theta).\text{ Also, } \theta = \arcsin(x)!$$

$$\therefore \int \sqrt{1-x^2}\, dx = \int \cos(\theta) \cdot \cos(\theta)\, d\theta = \int \cos^2(\theta)\, d\theta \overset{\boxed{\cos^2(\theta) = \frac{1+\cos(2\theta)}{2}}}{=} \frac{1}{2}\int 1 + \cos(2\theta)\, d\theta$$

$$= \frac{1}{2}\left(\theta + \frac{\sin(2\theta)}{2}\right) + C \overset{\boxed{\sin(2\theta) = 2\sin(\theta)\cos(\theta)}}{=} \frac{1}{2}\left(\arcsin(x) + \sin(\theta)\cos(\theta)\right) + C$$

From the diagram, $\cos(\theta) = \dfrac{\text{adjacent}}{\text{hypotenuse}} = \dfrac{\sqrt{1-x^2}}{1} = \sqrt{1-x^2}$.

$$\therefore \int \sqrt{1-x^2}\, dx = \frac{1}{2}\left(\arcsin(x) + x\sqrt{1-x^2}\right) + C$$

Example 2) Evaluate $\int \dfrac{x^2}{(9-4x^2)^{\frac{3}{2}}}\, dx$.

Solution In this example, $a = 3$ and $b = 2$. Let $2x = 3\sin(\theta)$. Then $dx = \dfrac{3}{2}\cos(\theta)\, d\theta$.

$$(9-4x^2)^{\frac{3}{2}} = (9-9\sin^2(\theta))^{\frac{3}{2}} = (9\cos^2(\theta))^{\frac{3}{2}} = (3|\cos(\theta)|)^3 \overset{\boxed{0<\theta<\frac{\pi}{2}}}{=} (3\cos(\theta))^3 = 27\cos^3(\theta).$$

Also, $\sin(\theta) = \dfrac{2x}{3}$ and $\theta = \arcsin\left(\dfrac{2x}{3}\right)$!

$$\therefore \int \dfrac{x^2}{(9-4x^2)^{\frac{3}{2}}}\, dx = \int \dfrac{\frac{9}{4}\sin^2(\theta)}{27\cos^3(\theta)} \cdot \frac{3}{2}\cos(\theta)\, d\theta \overset{\boxed{\text{This is the step where we hope the transformed integral has made life easier!}}}{=} \frac{1}{8}\int \dfrac{\sin^2(\theta)}{\cos^2(\theta)}\, d\theta$$

$$= \frac{1}{8}\int \tan^2(\theta)\, d\theta = \frac{1}{8}\int \sec^2(\theta) - 1\, d\theta = \frac{1}{8}\left(\tan(\theta) - \theta\right) + C$$

From the diagram at the right, $\tan(\theta) = \dfrac{\text{opposite}}{\text{adjacent}} = \dfrac{2x}{\sqrt{9-4x^2}}$

$$\therefore \int \dfrac{x^2}{(9-4x^2)^{\frac{3}{2}}}\, dx = \frac{1}{8}\left(\dfrac{2x}{\sqrt{9-4x^2}} - \arcsin\left(\dfrac{2x}{3}\right)\right) + C$$

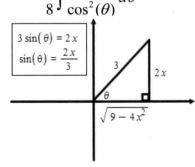

Two For You – Integration by Parts (I by P): $\int u\,dv = uv - \int v\,du$

1) Evaluate: $\int x\sin(x)\,dx$ 2) Evaluate: $\int x^2 e^x\,dx$

Answers   1) $\sin(x) - x\cos(x) + C$ 2) $x^2 e^x - 2xe^x + 2e^x + C$ $\overset{\text{or}}{=} e^x(x^2 - 2x + 2) + C$

Two For You – "Circular" Integration by Parts

1)(a) Redo $\int e^x \cos(x)\,dx$ but this time use $u = \cos(x)$ and $dv = e^x dx$ when you apply "I by P" first and then use $u = \sin(x)$ and $dv = e^x dx$ when you "I by P" again. Show that you get the same answer as in Example 1.

(b) Redo $\int e^x \cos(x)\,dx$ but this time use $u = \cos(x)$ and $dv = e^x dx$ when you "I by P" first and then use $u = e^x$ and $dv = \sin(x)\,dx$ when you "I by P" again. Show that you get $I = I$.

2) Evaluate: $\int e^x \sin(x)\,dx$

Answers   1)(a) $\dfrac{e^x}{2}(\sin(x) + \cos(x)) + C$ (b) $I = I$

2) $\dfrac{e^x}{2}(\sin(x) - \cos(x)) + C$

Two For You – Integration by Parts: The Tan—Sec Connection

1)(a) Finish Example 2: $\int \sec^5(x)\,dx$

(b) Evaluate: $\int \sec(x)\tan^2(x)\,dx$ (Hint: Let $u = \tan(x)$ and $dv = \sec(x)\tan(x)\,dx$.
Also, after doing I by P, use the substitution $\sec^2(x) = \tan^2(x) + 1$.)

2)(a) Evaluate: $\int \csc^3(x)\,dx$ (b) $\int \csc^5(x)\,dx$

> Hint: Special Toolbox for $\int \csc^3(x)\,dx$ and $\int \csc^5(x)\,dx$
>
> $\cot^2(x) = \csc^2(x) - 1$ $u = \csc(x) \Rightarrow du = -\csc(x)\cot(x)\,dx;$
>
> $\int \csc^2(x)\,dx = -\cot(x) + C$ $\int \csc(x)\,dx = \ln|\csc(x) - \cot(x)| + C$
>
> $u = \csc^3(x) = (\csc(x))^3 \Rightarrow du = 3(\csc(x))^2(-\csc(x)\cot(x))\,dx = -3\csc^3(x)\cot(x)\,dx$

Answers 1)(a) $\dfrac{1}{4}\sec^3(x)\cdot\tan(x) + \dfrac{3}{8}(\sec(x)\cdot\tan(x) + \ln|\sec(x) + \tan(x)|) + C$

(b) $\dfrac{1}{2}(\sec(x)\tan(x) - \ln|\sec(x) + \tan(x)|) + C$

2)(a) $\dfrac{1}{2}(\ln|\csc(x) - \cot(x)| - \csc(x)\cot(x)) + C$

(b) $\dfrac{3}{8}(\ln|\csc(x) - \cot(x)| - \csc(x)\cot(x)) - \dfrac{1}{4}\csc^3(x)\cot(x) + C$

Three For You – Integration by Trigonometric Substitution: Sin

1) Evaluate: $\displaystyle\int \dfrac{1}{(1-x^2)^{\frac{3}{2}}}\,dx$ 2) Evaluate: $\displaystyle\int \dfrac{\sqrt{25-16x^2}}{x^2}\,dx$

3) Why is it a bad idea to use trig substitution to evaluate $\int x\sqrt{1-x^2}\,dx$?

Answers 1) $\dfrac{x}{\sqrt{1-x^2}} + C$ 2) $-\dfrac{\sqrt{25-16x^2}}{x} - 4\arcsin\left(\dfrac{4x}{5}\right) + C$

3) It is so much easier to do this using the CHAIN RULE IN REVERSE!

$\displaystyle\int x\sqrt{1-x^2}\,dx = -\dfrac{1}{2}\int(-2x)(1-x^2)^{\frac{1}{2}}\,dx = \left(-\dfrac{1}{2}\right)\left(\dfrac{2}{3}\right)(1-x^2)^{\frac{3}{2}} + C = -\dfrac{1}{3}(1-x^2)^{\frac{3}{2}} + C$

Integration by Trigonometric Substitution: Tan

We use the tan substitution with the expression $a^2 + b^2x^2$, with a, b both positive. We set $bx = a\tan(\theta)$. Then $a^2 + b^2x^2 = a^2 + a^2\tan^2(\theta) = a^2(1 + \tan^2(\theta)) = a^2\sec^2(\theta)$.

With a little integration luck, a complicated algebraic integral will become a simple trigonometric integral. For simplicity, assume $0 < \theta < \frac{\pi}{2}$, ie., θ is in the first quadrant.

Example 1) Evaluate $\int \frac{1}{\sqrt{1+x^2}}\,dx$.

Solution Here, $a = 1$ and $b = 1$. Let $x = \tan(\theta)$. Then $dx = \sec^2(\theta)\,d\theta$.

$\sqrt{1+x^2} = \sqrt{1+\tan^2(\theta)} = \sqrt{\sec^2(\theta)} = |\sec(\theta)| \stackrel{0<\theta<\frac{\pi}{2}}{=} \sec(\theta)$. Also, $\theta = \arctan(x)$!

$\therefore \int \frac{1}{\sqrt{1+x^2}}\,dx = \int \frac{1}{\sec(\theta)}\sec^2(\theta)\,d\theta = \int \sec(\theta)\,d\theta = \ln|\sec(\theta) + \tan(\theta)| + C$

From the diagram, $\sec(\theta) = \frac{\text{hypotenuse}}{\text{adjacent}} = \sqrt{1+x^2}$.

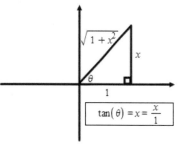

$\therefore \int \frac{1}{\sqrt{1+x^2}}\,dx = \ln|\sqrt{1+x^2} + x| + C$

(Fans of hyperbolic functions will recognize that $\ln|\sqrt{1+x^2} + x| = \operatorname{arctanh}(x)$!)

Example 2) Set up the transformed integral, using a tan substitution, for $\int \sqrt{16+9x^2}\,dx$.

Solution In this example, $a = 4$ and $b = 3$. Let $3x = 4\tan(\theta)$. Then $dx = \frac{4}{3}\sec^2(\theta)\,d\theta$.

$\sqrt{16+9x^2} = \sqrt{16+16\tan^2(\theta)} = \sqrt{16\sec^2(\theta)} = 4|\sec(\theta)| \stackrel{0<\theta<\frac{\pi}{2}}{=} 4\sec(\theta)$.

Also, $\tan(\theta) = \frac{3x}{4}$ and $\theta = \arctan\left(\frac{3x}{4}\right)$!

$\therefore \int \sqrt{16+9x^2}\,dx = \int 4\sec(\theta) \cdot \frac{4}{3}\sec^2(\theta)\,d\theta \stackrel{\text{This is the step where we hope the transformed integral has made life easier.}}{=} \frac{16}{3}\int \sec^3\,d\theta$

To finish this integration, use integration by parts. (See page 203)

Integration by Trigonometric Substitution: Sec

We use the sec substitution with the expression $b^2x^2 - a^2$, with a, b both positive. We set $bx = a\sec(\theta)$. Then $b^2x^2 - a^2 = a^2\sec^2(\theta) - a^2 = a^2(\sec^2(\theta) - 1) = a^2\tan^2(\theta)$.

With a little integration luck, a complicated algebraic integral will become a simple trigonometric integral. For simplicity, assume $0 < \theta < \dfrac{\pi}{2}$, ie., θ is in the first quadrant.

Example 1) Evaluate $\int x^3\sqrt{x^2 - 1}\,dx$.

Solution Here, $a = 1$ and $b = 1$. Let $x = \sec(\theta)$. Then $dx = \sec(\theta)\tan(\theta)\,d\theta$.

$\sqrt{x^2 - 1} = \sqrt{\sec^2(\theta) - 1} = \sqrt{\tan^2(\theta)} = |\tan(\theta)| \overset{0 < \theta < \frac{\pi}{2}}{=} \tan(\theta)$. Also, $\theta = \text{arcsec}(x)$!

$\therefore \int x^3\sqrt{x^2 - 1}\,dx = \int \sec^3(\theta)\tan(\theta) \cdot \sec(\theta)\tan(\theta)\,d\theta = \int \sec^4(\theta)\tan^2(\theta)\,d\theta$

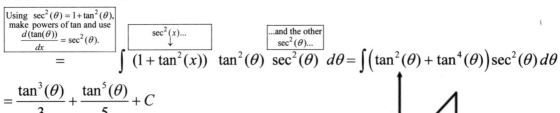

$= \int (1 + \tan^2(x))\ \tan^2(\theta)\ \sec^2(\theta)\,d\theta = \int \left(\tan^2(\theta) + \tan^4(\theta)\right)\sec^2(\theta)\,d\theta$

$= \dfrac{\tan^3(\theta)}{3} + \dfrac{\tan^5(\theta)}{5} + C$

From the diagram, $\tan(\theta) = \dfrac{\text{opposite}}{\text{adjacent}} = \sqrt{x^2 - 1}$.

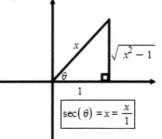

$\therefore \int x^3\sqrt{x^2 - 1}\,dx = \dfrac{(x^2-1)^{\frac{3}{2}}}{3} + \dfrac{(x^2-1)^{\frac{5}{2}}}{5} + C$

Example 2) Set up the transformed integral, using a sec substitution, for $\int \sqrt{x^2 - 16}\,dx$.

Solution In this example, $a = 4$ and $b = 1$. Let $x = 4\sec(\theta)$. Then $dx = 4\sec(\theta)\tan(\theta)\,d\theta$.

$\sqrt{x^2 - 16} = \sqrt{16\sec^2(\theta) - 16} = \sqrt{16\tan^2(\theta)} = 4|\tan(\theta)| \overset{0 < \theta < \frac{\pi}{2}}{=} 4\tan(\theta)$.

Also, $\sec(\theta) = \dfrac{x}{4}$ and $\theta = \arctan\left(\dfrac{x}{4}\right)$!

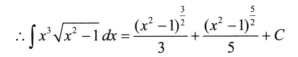

$\therefore \int \sqrt{x^2 - 16}\,dx = \int 4\tan(\theta) \cdot 4\sec(\theta)\tan(\theta)\,d\theta = 16\int \sec(\theta)\tan^2(\theta)\,d\theta$

To finish this integration, use integration by parts. (See page 203)

Integration Using Partial Fractions

In this section, let's take advantage of the work we did in the sections on partial fractions.

Example 1) Evaluate $\int \dfrac{x+1}{x^2-5x+4}\,dx$.

Solution From page 28, $\dfrac{x+1}{(x-4)(x-1)} = \dfrac{\tfrac{5}{3}}{x-4} - \dfrac{\tfrac{2}{3}}{x-1}$.

Remember that $\int \dfrac{1}{x\pm a}\,dx = \ln|x\pm a| + C$.

$$\therefore \int \dfrac{x+1}{x^2-5x+4}\,dx = \int \dfrac{\tfrac{5}{3}}{x-4} - \dfrac{\tfrac{2}{3}}{x-1}\,dx = \dfrac{5}{3}\ln|x-4| - \dfrac{2}{3}\ln|x-1| + C$$

Example 2) Evaluate $\int \dfrac{x^2+2x+1}{(x-1)^2(x+3)}\,dx$ in terms of partial fractions.

Solution From page 29, $\dfrac{x^2+2x+1}{(x-1)^2(x+3)} = \dfrac{1}{(x-1)^2} + \dfrac{\tfrac{3}{4}}{x-1} + \dfrac{\tfrac{1}{4}}{x+3}$.

$$\therefore \int \dfrac{x^2+2x+1}{(x-1)^2(x+3)}\,dx = \int \dfrac{1}{(x-1)^2} + \dfrac{\tfrac{3}{4}}{x-1} + \dfrac{\tfrac{1}{4}}{x+3}\,dx = \int (x-1)^{-2} + \dfrac{\tfrac{3}{4}}{x-1} + \dfrac{\tfrac{1}{4}}{x+3}\,dx$$

$$= -\dfrac{1}{x-1} + \dfrac{3}{4}\ln|x-1| + \dfrac{1}{4}\ln|x+3| + C$$

Example 3) Evaluate $\int \dfrac{x+2}{(x^2+1)(x+1)}\,dx$ in terms of partial fractions.

Solution From page 30, $\dfrac{x+2}{(x^2+1)(x+1)} = \dfrac{1}{2}\left(\dfrac{-x+3}{x^2+1} + \dfrac{1}{x+1}\right)$.

$$\therefore \int \dfrac{x+2}{(x^2+1)(x+1)}\,dx = \int \dfrac{1}{2}\left(\dfrac{-x+3}{x^2+1} + \dfrac{1}{x+1}\right)dx$$

Adjust the first term using $\int \dfrac{u'}{u}\,du = \ln|u| + C$.

$$= \dfrac{1}{2}\int -\dfrac{1}{2}\dfrac{(2x)}{x^2+1} + \dfrac{3}{x^2+1} + \dfrac{1}{x+1}\,dx$$

$$= \dfrac{1}{2}\left(-\dfrac{1}{2}\ln(x^2+1) + 3\arctan(x) + \ln|x+1|\right) + C$$

$$= -\dfrac{1}{4}\ln(x^2+1) + \dfrac{3}{2}\arctan(x) + \dfrac{1}{2}\ln|x+1| + C$$

Note x^2+1 is positive and so $\ln|x^2+1| = \ln(x^2+1)$.

Definite Integrals – Area Problems

Finding an area using definite integrals can easily be accomplished if you always follow the mantra "upper curve minus lower curve". You just need to keep in mind you need a new integral every time the curves intersect and the upper and lower curves switch.

Example 1) Find the area bounded by $y = x^2$ and $y = x + 2$.
Solution

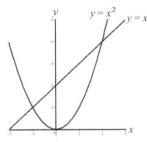

First we must find the x values of the points of intersection between the curve and the line.
$$x^2 = x + 2$$
$$x^2 - x - 2 = 0$$
$$(x-2)(x+1) = 0 \therefore x = 2 \text{ and } x = -1.$$

$$\text{Area} = \int_{x=-1}^{x=2} \text{upper curve} - \text{lower curve } dx$$

$$= \int_{x=-1}^{x=2} (x+2) - x^2 \, dx = \int_{x=-1}^{x=2} -x^2 + x + 2 \, dx = \left(-\frac{x^3}{3} + \frac{x^2}{2} + 2x \right) \Big|_{x=-1}^{x=2}$$

$$= \left(-\frac{2^3}{3} + \frac{2^2}{2} + 2(2) \right) - \left(-\frac{(-1)^3}{3} + \frac{(-1)^2}{2} + 2(-1) \right) = \frac{9}{2} \text{ units}^2$$

Example 2) Find the area between $y = x^3 - x$ and the x axis from $x = -1$ to $x = 1$.
Solution

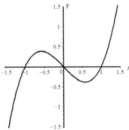

The two curves in this question are $y = x^3 - x$ and $y = 0$ (the x axis). From $x = -1$ to $x = 0$, the cubic is the upper curve and the x axis is the lower curve. From $x = 0$ to $x = 1$, the axis is the upper curve and the cubic is the lower curve. This means we will need two integrals.

$$\text{Area} = \int_{x=-1}^{x=0} \text{upper curve} - \text{lower curve } dx + \int_{x=0}^{x=1} \text{upper curve} - \text{lower curve } dx$$

$$= \int_{x=-1}^{x=0} (x^3 - x) - 0 \, dx + \int_{x=0}^{x=1} 0 - (x^3 - x) \, dx = \int_{x=-1}^{x=0} x^3 - x \, dx + \int_{x=0}^{x=1} -x^3 + x \, dx$$

$$= \left(\frac{x^4}{4} - \frac{x^2}{2} \right) \Big|_{x=-1}^{x=0} + \left(-\frac{x^4}{4} + \frac{x^2}{2} \right) \Big|_{x=0}^{x=1} = \left[\left(\frac{0}{4} - \frac{0}{2} \right) - \left(\frac{1}{4} - \frac{1}{2} \right) \right] + \left[\left(-\frac{1}{4} + \frac{1}{2} \right) - \left(-\frac{0}{4} + \frac{0}{2} \right) \right] = \frac{1}{2} \text{ units}^2$$

Three For You – Integration by Trigonometric Substitution: Tan

1) Evaluate: $\int \dfrac{x^3}{\sqrt{1+x^2}}\,dx$

2) Evaluate: $\int \dfrac{x+1}{\sqrt{25+16x^2}}\,dx$

3) Why is it a bad idea to use trig substitution to evaluate $\int x\sqrt{1+x^2}\,dx$?

Answers 1) $\dfrac{(1+x^2)^{\frac{3}{2}}}{3} - \sqrt{1+x^2} + C$

2) $\dfrac{\sqrt{25+16x^2}}{16} + \dfrac{1}{4}\ln\left(4x+\sqrt{25+16x^2}\right) + C$

3) It is so much easier to do this using the CHAIN RULE IN REVERSE!

$\int x\sqrt{1+x^2}\,dx = \dfrac{1}{2}\int (2x)(1+x^2)^{\frac{1}{2}}\,dx = \left(\dfrac{1}{2}\right)\left(\dfrac{2}{3}\right)(1+x^2)^{\frac{3}{2}} + C = \dfrac{1}{3}(1+x^2)^{\frac{3}{2}} + C$

Three For You – Integration by Trigonometric Substitution: Sec

1) Evaluate: $\int \dfrac{x^3}{\sqrt{x^2-1}}\,dx$

2) Evaluate: $\int \dfrac{1}{\sqrt{4x^2-9}}\,dx$

3) Why is it a bad idea to use trig substitution to evaluate $\int x\sqrt{x^2-1}\,dx$?

Answers 1) $\dfrac{(x^2-1)^{\frac{3}{2}}}{3} + \sqrt{x^2-1} + C$

2) $\dfrac{1}{2}\ln|2x+\sqrt{4x^2-9}| + C$

3) It is so much easier to do this using the CHAIN RULE IN REVERSE!

$\int x\sqrt{x^2-1}\,dx = \dfrac{1}{2}\int (2x)(x^2-1)^{\frac{1}{2}}\,dx = \left(\dfrac{1}{2}\right)\left(\dfrac{2}{3}\right)(x^2-1)^{\frac{3}{2}} + C = \dfrac{1}{3}(x-1^2)^{\frac{3}{2}} + C$

Three For You – Integration Using Partial Fractions

Evaluate:

1) $\int \dfrac{x+5}{x^3 - x} dx.$ 2) $\int \dfrac{2x-1}{(x+1)^2 (x-1)} dx$ 3) $\int \dfrac{x-3}{(x^2+1)x^2} dx$

Hint: You already did the work to rewrite the integrands in terms of partial fractions on pages 31 and 32.

Answers 1) $2\ln|x+1| + 3\ln|x-1| - 5\ln|x| + C$

2) $-\dfrac{3}{2(x+1)} - \dfrac{1}{4}\ln|x+1| + \dfrac{1}{4}\ln|x-1| + C$

3) $\dfrac{3}{x} - \dfrac{1}{2}\ln(x^2+1) + 3\arctan(x) + \ln|x| + C$

Two You – Definite Integrals – Area Problems

1) Find the area bounded by the y axis, $y = x^3$ and $y = 2x + 4$.
2) Find the area bounded by $y = \sin(x)$ and $y = \cos(x)$ between $x = 0$ and $x = \pi$.

Hint: Start by sketching the two curves and finding the points of intersection.

Answers 1) Area $= \displaystyle\int_{x=0}^{x=2} (2x+4) - x^3 \, dx = 8$ units^2

2) Area $= \displaystyle\int_{x=0}^{x=\pi/4} \cos(x) - \sin(x)\, dx + \int_{x=\pi/4}^{x=\pi} \sin(x) - \cos(x)\, dx = \dfrac{4}{\sqrt{2}} = 2\sqrt{2}$ units^2

Definite Integrals Using Substitution

When solving an integral in the variable x using a substitution, we transform the x expression and dx to the new variable. If it is a **definite** integral, we change the limits of integration too.

Example 1) Evaluate $\int_{x=-1}^{0} (x-1)\sqrt{x+1}\, dx$.

Solution This integral looks a little scary, but is easily tackled using a substitution. Let $t = x+1$. This means $x = t-1$ and therefore $dx = dt$. In the original integral, x goes from -1 to 0. Since $t = x+1$, t must go from 0 to 1.

$$\therefore \int_{x=-1}^{0} (x-1)\sqrt{x+1}\, dx = \int_{t=0}^{1} (t-1-1)\sqrt{t}\, dt$$

$$= \int_{t=0}^{1} (t-2)\left(t^{1/2}\right) dt = \int_{0}^{1} t^{\frac{3}{2}} - 2t^{\frac{1}{2}}\, dt = \left(\frac{2}{5}t^{\frac{5}{2}} - \frac{4}{3}t^{\frac{3}{2}}\right)\Big|_{t=0}^{t=1}$$

$$= \left(\frac{2}{5}(1^{5/2}) - \frac{4}{3}\left(1^{\frac{3}{2}}\right)\right) - 0 = -\frac{14}{15}$$

Example 2) Evaluate $\int_{x=0}^{4} \frac{x}{\sqrt{2x+1}}\, dx$.

Solution Let $t = 2x+1$. This means $x = \frac{1}{2}(t-1)$ and therefore $dx = \frac{1}{2}dt$. In the original integral, x goes from 0 to 4. Since $t = 2x+1$, t must go from 1 to 9.

$$\therefore \int_{x=0}^{4} \frac{x}{\sqrt{2x+1}}\, dx = \int_{t=1}^{9} \frac{\frac{1}{2}(t-1)}{\sqrt{t}}\left(\frac{1}{2}\right) dt$$

$$= \frac{1}{4}\int_{t=1}^{9} \frac{t-1}{t^{1/2}}\, dt \quad \boxed{\text{Make separate fractions.}} \quad = \frac{1}{4}\int_{t=1}^{9} \frac{t}{t^{1/2}} - \frac{1}{t^{1/2}}\, dt = \frac{1}{4}\int_{t=1}^{9} t^{1/2} - t^{-1/2}\, dt$$

$$= \frac{1}{4}\left(\frac{2}{3}t^{3/2} - \frac{2}{1}t^{1/2}\right)\Big|_{t=1}^{t=9}$$

$$= \frac{1}{4}\left[\left(\frac{2}{3}(9^{3/2}) - 2(9^{1/2})\right) - \left(\frac{2}{3}(1^{3/2}) - 2(1^{1/2})\right)\right] = \frac{10}{3}$$

Improper Integrals – Functions with a Discontinuity

First, a reminder. If $f(x) \geq 0$, $\int_{x=a}^{b} f(x)\,dx$ measures the area between $f(x)$ and the x axis from $x = a$ to $x = b$. Here is a question that contains an **improper integral**: $\int_{x=0}^{2} \frac{1}{(x-1)^2}\,dx$. The function has a vertical asymptote at $x = 1$ and we are integrating from $x = 0$ to 2. Let's ignore the discontinuity and integrate:

$$\int_{x=0}^{2} \frac{1}{(x-1)^2}\,dx = \int_{x=0}^{2} (x-1)^{-2}\,dx = -(x-1)^{-1}\Big|_{x=0}^{x=2} = -\frac{1}{1} - \left(-\frac{1}{-1}\right) = -2,\text{ a negative area!}$$

But $\frac{1}{(x-1)^2} > 0$, and so the integral must be positive! How can we resolve this seeming contradiction? We must carefully set up limits so that we integrate everything in the interval except for the problem at $x = 1$.

Example 1) Evaluate $\int_{x=0}^{2} \frac{1}{(x-1)^2}\,dx$. (Correctly this time!)

Solution

$$\int_{x=0}^{2} \frac{1}{(x-1)^2}\,dx = \lim_{a \to 1^-} \int_{x=0}^{a} \frac{1}{(x-1)^2}\,dx + \lim_{b \to 1^+} \int_{x=b}^{2} \frac{1}{(x-1)^2}\,dx$$

$$= \lim_{a \to 1^-} \left(-(x-1)^{-1}\Big|_{x=0}^{x=a}\right) + \lim_{b \to 1^+} \left(-(x-1)^{-1}\Big|_{x=b}^{x=2}\right)$$

$$= \lim_{a \to 1^-} \left(-\frac{1}{a-1} + 1\right) + \lim_{b \to 1^+} \left(-1 + \frac{1}{b-1}\right)$$

$$\overset{"(\infty+1)+(-1+\infty)"}{=} \infty$$

Example 2) Evaluate $\int_{x=0}^{1} \frac{1}{(x-1)^{2/3}}\,dx$.

Solution

$$\int_{x=0}^{1} \frac{1}{(x-1)^{2/3}}\,dx \underset{\text{Note the vertical asymptote at }x=1.}{=} \lim_{a \to 1^-} \int_{x=0}^{a} (x-1)^{-2/3}\,dx$$

$$= \lim_{a \to 1^-} \left(3(x-1)^{1/3}\right)\Big|_{x=0}^{x=a}$$

$$= \lim_{a \to 1^-} \left(3(a-1)^{1/3} - 3(0-1)^{1/3}\right)$$

$$= 0 - 3(-1)^{1/3}$$

$$= 3$$

Improper Integrals – Infinite Limits of Integration

When you look at the integral $\int_{x=0}^{\infty} \frac{1}{(x+1)^2} \, dx$, you may think "I'll just do the integral the usual way and then plug in infinity at the end!" However, every time you do that your first year calculus professor has an anxiety attack of infinite proportions. You can run into all kinds of numerical paradoxes if you plug in ∞ and treat it as an ordinary number! However, you **can** plug in a variable and let that variable approach infinity in a limit.

Example 1) Evaluate $\int_{x=0}^{\infty} \frac{1}{(x+1)^2} \, dx$.

Solution

$$\int_{x=0}^{\infty} \frac{1}{(x+1)^2} \, dx = \lim_{a \to \infty} \int_{x=0}^{a} \frac{1}{(x+1)^2} \, dx$$

$$= \lim_{a \to \infty} \int_{x=0}^{a} (x+1)^{-2} \, dx$$

$$= \lim_{a \to \infty} \left(-(x+1)^{-1} \right) \Big|_{x=0}^{x=a}$$

$$= -\lim_{a \to \infty} \left(\frac{1}{a+1} - \frac{1}{0+1} \right)$$

$$= -(0-1) = 1$$

Example 2) Evaluate $\int_{-\infty}^{\infty} \frac{1}{1+x^2} \, dx$.

Solution

$\int_{-\infty}^{\infty} \frac{1}{1+x^2} \, dx$ [You must use a different letter for each infinity.] $= \lim_{\substack{a \to -\infty \\ b \to \infty}} \int_{x=a}^{b} \frac{1}{1+x^2} \, dx$

[Note that $\int \frac{1}{1+x^2} \, dx = \arctan(x) + C$.]

$$= \lim_{\substack{a \to -\infty \\ b \to \infty}} \left(\arctan(x) \right) \Big|_{x=a}^{x=b}$$

$$= \lim_{\substack{a \to -\infty \\ b \to \infty}} \left(\arctan(b) - \arctan(a) \right)$$

[See the graph of arctan on page 141.]

$$= \frac{\pi}{2} - \left(-\frac{\pi}{2} \right)$$

$$= \pi$$

The Derivative of an Integral

This is an application of: **The Fundamental Theorem of Calculus**. In this application, we take the derivative of an integral. You might expect the answer to be the original question. It is. Almost.

Example 1) Evaluate: (a) $\dfrac{d}{dx}\left(\displaystyle\int_{t=1}^{x} t^3 + 1 \, dt\right)$ (b) $\dfrac{d}{dx}\left(\displaystyle\int_{t=1}^{x} t^3 + t + 5 \, dt\right)$

Solution (a) $\dfrac{d}{dx}\left(\displaystyle\int_{t=1}^{x} t^3 + 1 \, dt\right) = \dfrac{d}{dx}\left(\dfrac{t^4}{4} + t\right)\Big|_{1}^{x} = \dfrac{d}{dx}\left(\dfrac{x^4}{4} + x - \left(\dfrac{1}{4} + 1\right)\right) = x^3 + 1$

We are back where we started except that t is replaced by x! Knowing this happens...

(b) $\dfrac{d}{dx}\left(\displaystyle\int_{t=1}^{x} t^3 + t + 5 \, dt\right) \overset{\text{The answer is the original integrand with } t \text{ replaced by } x!}{=} x^3 + x + 5$

Example 2) (a) $\dfrac{d}{dx}\left(\displaystyle\int_{t=\sin(x)}^{\tan(x)} t^3 + 1 \, dt\right)$ (b) $\dfrac{d}{dx}\left(\displaystyle\int_{t=B(x)}^{T(x)} f(t) \, dt\right)$ $\boxed{\begin{array}{l}T \text{ for } Top\\ B \text{ for } Bottom\end{array}}$

Solution (a) $\dfrac{d}{dx}\left(\displaystyle\int_{t=\sin(x)}^{\tan(x)} t^3 + 1 \, dt\right) \overset{\text{This is just like Example 1(a) so far.}}{=} \dfrac{d}{dx}\left(\dfrac{t^4}{4} + t\right)\Big|_{\sin(x)}^{\tan(x)}$

$= \dfrac{d}{dx}\left(\dfrac{\tan^4(x)}{4} + \tan(x) - \left(\dfrac{\sin^4(x)}{4} + \sin(x)\right)\right)$

$\boxed{\text{NOW WE NEED THE CHAIN RULE.}}$

$= \tan^3(x)\sec^2(x) + \sec^2(x) - \left(\sin^3(x)\cos(x) + \cos(x)\right)$

$\boxed{\text{Factor out the "}\sec^2(x)\text{" and "}\cos(x)\text{" terms.}}$

$= (\tan^3(x) + 1)\sec^2(x) - \left(\sin^3(x) + 1\right)\cos(x)$

Here, we come back to the original integrand, but "t" is replaced **first** with "$\tan(x)$" and the chain rule made us multiply by $\dfrac{d\tan(x)}{dx} = \sec^2(x)$, then by "$\sin(x)$" and the chain rule made us multiply by $\dfrac{d\sin(x)}{dx} = \cos(x)$. So, knowing that this is what happens...

(b) $\dfrac{d}{dx}\left(\displaystyle\int_{t=B(x)}^{T(x)} f(t) \, dt\right) = f(T(x))\dfrac{d(T(x))}{dx} - f(B(x))\dfrac{d(B(x))}{dx} \overset{\text{in short...}}{=} f(T)T' - f(B)B'$

Two For You – Definite Integrals Using Substitution

Evaluate the following integrals:

1) $\displaystyle\int_{x=-1}^{7} (7x+3)(x+1)^{1/3}\,dx$

2) $\displaystyle\int_{x=0}^{4} \frac{x+x^2}{\frac{x}{2}+1}\,dx$

Answers 1) 336   2) $8+4\ln(3)$

Two For You – Improper Integrals – Functions with a Discontinuity

Evaluate the following integrals:

1)(a) $\displaystyle\int_{x=8}^{10} \frac{1}{(x-9)^2}\,dx$ (b) $\displaystyle\int_{x=8}^{10} \frac{1}{(x-9)^{\frac{2}{3}}}\,dx$

2) $\displaystyle\int_{x=0}^{\frac{\pi}{2}} \sec(x)\,dx$ Recall: $\int \sec(x)\,dx = \ln(|\sec(x)+\tan(x)|)+C$

Answers 1)(a) ∞ (b) 6   2) ∞

Two For You – Improper Integrals – Infinite Limits of Integration

Evaluate the following integrals:

1)(a) $\displaystyle\int_{-\infty}^{-1} \frac{1}{x^2}\,dx$ (b) $\displaystyle\int_{x=1}^{\infty} \frac{1}{x}\,dx$

2) $\displaystyle\int_{-\infty}^{\infty} xe^{-x^2}\,dx$

Answers 1)(a) 1   (b) ∞ 2) 0

Two For You – The Derivative of an Integral

Evaluate the following:

1) $\displaystyle\frac{d}{dx}\left(\int_{t=e^x}^{x^2+1} \cos t\,\ln t + t\,dt\right)$ (Hint: remember $\ln(e^x) = x$.)   2) $\displaystyle\frac{d}{dx}\left(\int_{\ln x}^{x} x + \ln x\,dx\right)$

Answers 1) $2x\left(\cos(x^2+1)\,\ln(x^2+1) + x^2+1\right) - e^x\left(x\cos(e^x) + e^x\right)$

2) $x + \ln x - \dfrac{\ln x + \ln(\ln x)}{x}$

(Note: the variable in the integrand, "x" or "t", **does not matter!**)

Differential Equations – Separation of Variables

It is very easy to solve the equation $y' = 2x$. By integrating, we find $y = x^2 + C$, where C is some constant. There is a whole branch of mathematics dedicated to studying equations involving multiple levels of derivatives, for example: "Given $y'' - y' + y = 2x$, solve for y". If you take a course in differential equations, you will learn more in this area. We will focus on one of the basic solution techniques, **separation of variables**. This technique works for differential equations of the form $y' = g(y)f(x)$, such as $y' = e^{-y}x$.

Example 1) Find the **general solution** to $y' = e^{-y}x$ and the **particular solution** that passes through the point $(0,0)$.

Solution With a slight notation change, the equation is: $\dfrac{dy}{dx} = e^{-y}x \Rightarrow e^y \dfrac{dy}{dx} = x$.

Now integrate each side with respect to x.

$\int e^y \dfrac{dy}{dx} dx = \int x\, dx$ [Think of $\dfrac{dy}{dx}$ as a fraction. This is actually the chain rule.] $\Rightarrow \int e^y dy = \int x\, dx$ [Note: The arbitrary constant from the integration on the left side can be absorbed into the one on the right side.] $\Rightarrow e^y = \dfrac{x^2}{2} + C$

$\therefore y = \ln\left(\dfrac{x^2}{2} + C\right)$. This is the general solution since it still has the constant $+C$.

To find a particular solution, you need more information in order to solve for C. In our case, we are told our solution passes through $(0,0)$. Plugging this information in gives $0 = \ln(0 + C) \therefore C = 1$. This means our particular solution is $y = \ln\left(\dfrac{x^2}{2} + 1\right)$.

Example 2) Find the general solution to $\dfrac{dy}{dx} = y\sin(x)$.

Solution

$\dfrac{1}{y}\dfrac{dy}{dx} = \sin(x)$ [Integrate each side with respect to x.] $\Rightarrow \int \dfrac{1}{y}\dfrac{dy}{dx} dx = \int \sin(x)\, dx \Rightarrow \int \dfrac{1}{y} dy = -\cos(x) + C$

$\therefore \ln|y| = -\cos(x) + C \Rightarrow y = e^{-\cos(x) + C} = e^{-\cos(x)} e^C$ [Let $A = e^C$.] $= Ae^{-\cos(x)}$.

Finding the Inverse of a Function

Let $f(x) = x^3 + 1$. Then $f(2) = 9$. In the inverse function, we should have $f^{-1}(9) = 2$. In other words, since the point $(2, 9)$ satisfies f, the point $(9, 2)$ must satisfy f^{-1}. The key to inverses is that **the roles of x and y are INTERCHANGED!** So to find the inverse of a function, we 1) **interchange x and y** and then 2) **solve for y.**

Example 1)(a) Find the inverse function for $f(x) = x^3 + 1$.

(b) Verify that $f^{-1}(f(x)) = x$ for $x \in \text{dom}(f)$ and $f(f^{-1}(x)) = x$ for $x \in \text{dom}(f^{-1})$.

(c) In general, how are the domains and ranges of f and f^{-1} related?

(d) Draw the graphs of f, f^{-1}, and $y = x$ on the same set of axes. How are the three graphs related?

Solution (a) First, write $y = x^3 + 1$.

Interchange x and y: $x = y^3 + 1$.

Now, solve for y: $y^3 = x - 1$ and so $y = f^{-1}(x) = (x-1)^{1/3}$. Note $f^{-1}(9) = (9-1)^{1/3} = 8^{1/3} = 2$. It works!

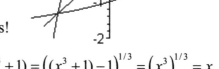

(b) $x \in \text{dom}(f)$: $f^{-1}(f(x)) = f^{-1}(x^3 + 1) = ((x^3 + 1) - 1)^{1/3} = (x^3)^{1/3} = x$

$x \in \text{dom}(f^{-1})$: $f(f^{-1}(x)) = f((x-1)^{1/3}) = ((x-1)^{1/3})^3 + 1 = x - 1 + 1 = x$

(c) $\text{domain}(f) = \text{range}(f^{-1})$ $\text{range}(f) = \text{domain}(f^{-1})$

(d) f and f^{-1} are mirror images in the line $y = x$!
For example, look at points $(1, 0)$ and $(0, 1)$.

> **Note**: given the function $y = f(x)$, the inverse will be a function only if f is **one to one**! For example, if $f(x) = x^2$, then the inverse **relation** is $f^{-1}(x) = \pm\sqrt{x}$. Since $f(3) = f(-3) = 9$, therefore $f^{-1}(9) = \pm 3$; f^{-1} is **not** a function!

Example 2)(a) Find the inverse of $y = f(x) = \ln(x+1)$.

(b) State the domain and range for each of f and f^{-1}.

Solution (a) Set $x = \ln(y+1)$. $\therefore y + 1 = e^x$ and so $y = f^{-1}(x) = e^x - 1$.

(b) Domain of f: $x + 1 > 0$ and so $x > -1$. Range of f: $y \in \mathbb{R}$.

Domain of f^{-1}: $x \in \mathbb{R}$. Range of f^{-1}: $y = e^x - 1 > -1$.

The domains and ranges of f and f^{-1} are interchanged!

Derivatives of Inverse Functions

Let $y = f(x)$. Then $\dfrac{dy}{dx} = f'(x)$. For the inverse, we want x and y to switch roles, so the derivative of the inverse should be, at least in notation, $\dfrac{dx}{dy}$. But remember that we can treat dx and dy as separate quantities (run and rise along the tangent line – see **Estimating Using the Differential** on page 178). So, from Grade 5 arithmetic,

$\dfrac{dx}{dy} \overset{\text{should}}{=} \dfrac{1}{\left(\dfrac{dy}{dx}\right)}$. **AND IT DOES!** However, there is a **subtle** part: we have switched x and y. We evaluate $\dfrac{dx}{dy}$ at the point (y, x) while we evaluate $\dfrac{dy}{dx}$ at (x, y).

> The derivative of the inverse function at the point (y, x) is the reciprocal of the derivative of the original function at the point (x, y).

Example 1) Let $y = f(x) = x^3 + 1$. (a) Find $f^{-1}(x)$, $f'(x)$, and $(f^{-1})'(x)$.
(b) Note that $f(2) = 9$. Verify that $(f^{-1})'(9) = \dfrac{1}{f'(2)}$.

Solution (a) In $y = f(x) = x^3 + 1$, interchange x and y: $x = y^3 + 1$. Now solving for y, $y^3 = x - 1$ and so $y = f^{-1}(x) = (x-1)^{1/3}$.
Therefore, $f'(x) \overset{\boxed{f(x)=x^3+1}}{=} 3x^2$ and $(f^{-1})'(x) \overset{\boxed{f^{-1}(x)=(x-1)^{1/3}}}{=} \dfrac{1}{3(x-1)^{2/3}}$.

(b) $f'(2) = 12$ and $(f^{-1})'(9) = \dfrac{1}{3(8)^{2/3}} = \dfrac{1}{12} = \dfrac{1}{f'(2)}$

Example 2) Suppose $f(7) = 8$ and $f'(7) = \dfrac{4}{5}$. Find (i) $(f^{-1})'(8)$ (ii) $(f^{-1})'(7)$

Solution (i) $(f^{-1})'(8) = \dfrac{1}{f'(7)} = \dfrac{5}{4}$ (ii) You don't have enough information!

Parametric Equations

There is a problem with finding specific points for the circle $x^2 + y^2 = 1$. If we set $y = 0$, we get two x values: $x = \pm 1$. Can we find another equation for the circle where substituting a value for a variable yields exactly one of these x values? There is a neat way to do this. We introduce a "parameter" t and find two equations for x and y in terms of t that (i) plot the circle and (ii) for each t, we get exactly one point (x,y) on the circle.

Example 1) Find a pair of parametric equations for the circle $x^2 + y^2 = 1$ that plots the circle starting at $(1,0)$ and tracing counter-clockwise to the point $(-1,0)$.

Solution Let's take advantage of the famous trigonometric equation: $\cos^2(t) + \sin^2(t) = 1$. Let $x = \cos(t)$ and $y = \sin(t)$ and choose $0 \le t \le \pi$. Since $x^2 + y^2 = \cos^2(t) + \sin^2(t) = 1$, each value of t gives a point $(\cos(t), \sin(t))$ on the circle. Also, each point on the semi-circle from $(1,0)$ to $(-1,0)$ is given by exactly one value of t in the interval $[0, \pi]$.

Table of Values

t	0	$\pi/2$	π
x	1	0	-1
y	0	1	0

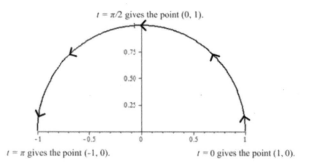

Note that the parametric equations allow us to define a portion of the circle and the direction in which to trace it. This is **much more useful** than function notation. We pay a price for this flexibility, which I will mention after Example 2.*

Example 2) Find two pairs of parametric equations for the circle $x^2 + y^2 = 9$ that plot the circle starting at $(0,-3)$ and tracing clockwise to the point $(3,0)$.

Solution Let's again take advantage of the famous trigonometric equation: $9\cos^2(t) + 9\sin^2(t) = 9$. (Note that the 9 ensures the radius of the circle is 3.)
Solution 1) Let $x = 3\cos(t)$ and $y = 3\sin(t)$ and choose $-\pi/2 \le t \le \pi$.
Solution 2) Let $x = 3\sin(t)$ and $y = 3\cos(t)$ and choose $-\pi \le t \le \pi/2$.

*There are an infinite number of ways to "parameterize" a curve. By comparison, there is only one way to write a function y in terms of x.

Two For You – Differential Equations – Separation of Variables

1) Find the general solution to $y' = 3y^2 x^2$ using separation of variables. Find the particular solution that passes through the point $\left(0, \dfrac{1}{2}\right)$.

2) Find the general solution to $y' = \dfrac{1+y^2}{1+x^2}$ using separation of variables. Find the particular solution that passes through the point $(0,0)$. Hint: $\int \dfrac{1}{1+t^2}\, dt = \arctan(t) + C$

Answers 1) general solution: $y = -\dfrac{1}{x^3 + C}$; particular solution: $C = -2$ and so, $y = -\dfrac{1}{x^3 - 2}$

2) general solution: $y = \tan(\arctan(x) + C)$; particular solution: $C = 0$ and so, $y = x$

Two For You – Finding the Inverse of a Function

Find f^{-1} for each of the following: 1) $f(x) = x^5 + 5$ 2) $f(x) = e^{2x+1}$

Answers 1) $f^{-1}(x) = (x-5)^{1/5}$ 2) $f^{-1}(x) = \dfrac{\ln x - 1}{2}$

Two For You – Derivatives of Inverse Functions

For the function $y = f(x) = x^2 + 4$, find $(f^{-1})'$ at each of these points of f:
1) $(2, 8)$ and 2) $(-2, 8)$

Answers 1) $\dfrac{1}{4}$ 2) $-\dfrac{1}{4}$

(Note that there would be a problem asking for "$(f^{-1})'(8)$"! Since $f(x)$ is not a one to one function, f^{-1} is **NOT A FUNCTION!** For example, $f(2) = f(-2) = 8$. We **can't** have $(f^{-1})'(8) = \dfrac{1}{f'(2)}$ and $(f^{-1})'(8) = \dfrac{1}{f'(-2)}$.)

Three For You – Parametric Equations

1) Find two pairs of parametric equations for the circle $x^2 + y^2 = 25$ that plot starting at $(0,5)$ trace **counter-clockwise** to $(5,0)$.

2) Find a pair of parametric equations for the ellipse $4x^2 + 9y^2 = 36$ that plots starting at $(-3, 0)$ and traces **clockwise** to $(0, 2)$.

3) Find a pair of parametric equations for the parabola $y^2 = x$ that plots in the direction of **decreasing** y from $(1,1)$ to $(1,-1)$.

Answers
1) $x = 5\cos(\theta),\ y = 5\sin(\theta),\ \theta \in \left[\dfrac{\pi}{2}, 2\pi\right]$; $x = -5\sin(\theta),\ y = 5\cos(\theta),\ \theta \in \left[0, \dfrac{3\pi}{2}\right]$

2) $2x = 6\cos(\theta)$ so $x = 3\cos(\theta),\ 3y = -6\sin(\theta)$ so $y = -2\sin(\theta),\ \theta \in \left[\pi, \dfrac{3\pi}{2}\right]$

3) $x = t^2,\ y = -t,\ t \in [-1, 1]$

In all three questions, there are an infinite number of correct answers!

Derivatives from Parametric Equations

Given $y = f(t)$ and $x = g(t)$, what is $\dfrac{dy}{dx}$? Easy, as long as we remember

(i) the Chain Rule and (ii) $\dfrac{dt}{dx} = \dfrac{1}{\left(\dfrac{dx}{dt}\right)}$. If we think of dx and dt as non-zero values.

(see The Differential, page 178), this is just our "invert and multiply" fraction rule.

$$\dfrac{dy}{dx} \underset{\substack{\text{The Chain Rule}\\ \text{First take the derivative}\\ \text{with respect to } t.}}{=} \dfrac{dy}{dt} \cdot \dfrac{dt}{dx} \underset{\frac{dt}{dx} = \frac{1}{\left(\frac{dx}{dt}\right)}}{=} \dfrac{\left(\dfrac{dy}{dt}\right)}{\left(\dfrac{dx}{dt}\right)} = \dfrac{f'(t)}{g'(t)}$$

Example 1(a) Find $\dfrac{dy}{dx}$ if $y = t^3 - 3t$ and $x = t^2$.

(b) if $w = \cos(s)$ and $z = \sin(s)$, find (i) $\dfrac{dw}{dz}$ (ii) $\dfrac{dz}{dw}$.

Solution (a) $\dfrac{dy}{dx} = \dfrac{\left(\dfrac{dy}{dt}\right)}{\left(\dfrac{dx}{dt}\right)} = \dfrac{3t^2 - 3}{2t}$

(b)(i) $\dfrac{dw}{dz} = \dfrac{\left(\dfrac{dw}{ds}\right)}{\left(\dfrac{dz}{ds}\right)} = \dfrac{-\sin(s)}{\cos(s)} = -\tan(s)$ (ii) $\dfrac{dz}{dw} = \dfrac{1}{\left(\dfrac{dw}{dz}\right)} = \dfrac{1}{-\tan(s)} = -\cot(s)$

Note that $\dfrac{dw}{dz}$ and $\dfrac{dz}{dw}$ are reciprocals!

Example 2) (a) Find the t values that give (i) horizontal (ii) vertical tangents to the the curve in the xy plane defined by $y = t^3 - 3t$ and $x = t^2$.

Solution Note that x and y are defined for $t \in \mathbb{R}$.

$$\dfrac{dy}{dx} = \dfrac{3t^2 - 3}{2t} = \dfrac{3(t-1)(t+1)}{2t}$$

(i) Horizontal tangent (slope = 0): $\dfrac{dy}{dx} = 0 \Rightarrow t = -1$ and $t = 1$.

(ii) Vertical Tangent (infinite slope): $\dfrac{dy}{dx}$ has infinite slope $\Rightarrow t = 0$.

Two subtleties to note. First, t must be in the domain of f and g. Otherwise, we may be dealing with an asymptote (see Vertical and Horizontal Asymptotes, page 87), not a tangent. Second, if both dy/dt and dx/dt equal 0 for the same t value, we need to do a little more work to determine if the tangent is vertical or horizontal, if either.

Higher Derivatives from Parametric Equations

Given $y = f(t)$ and $x = g(t)$, what is $\dfrac{d^2y}{dx^2}$? We know that $\dfrac{dy}{dx} = \dfrac{\left(\dfrac{dy}{dt}\right)}{\left(\dfrac{dx}{dt}\right)} = \dfrac{f'(t)}{g'(t)}$.

Let's write $\dfrac{dy}{dx} = w$. Then

$$\dfrac{d^2y}{dx^2} \underset{\text{Take the derivative with respect to } x \text{ of } \frac{dy}{dx}}{=} \dfrac{d}{dx}\left(\dfrac{dy}{dx}\right) \underset{\text{...which is the same as ...}}{=} \dfrac{d\left(\dfrac{dy}{dx}\right)}{dx} \underset{\frac{dy}{dx}=w}{=} \dfrac{dw}{dx}.$$

Now find $\dfrac{dw}{dx}$ exactly as we found $\dfrac{dy}{dx}$.

$$\dfrac{dw}{dx} = \dfrac{\left(\dfrac{dw}{dt}\right)}{\left(\dfrac{dx}{dt}\right)} \underset{w=\frac{dy}{dx}}{=} \dfrac{\dfrac{d}{dt}\left(\dfrac{dy}{dx}\right)}{\left(\dfrac{dx}{dt}\right)}, \text{ that is, } \dfrac{d^2y}{dx^2} = \dfrac{\dfrac{d}{dt}\left(\dfrac{dy}{dx}\right)}{\left(\dfrac{dx}{dt}\right)} = \dfrac{\dfrac{d}{dt}\left(\dfrac{f'(t)}{g'(t)}\right)}{g'(t)}.$$

Example 1) (a) Find $\dfrac{d^2y}{dx^2}$ if $y = t^3 - 3t$ and $x = t^2$.

(b) Find $\dfrac{d^2w}{dz^2}$ if $w = \cos(s)$ and $z = \sin(s)$.

Solution (a) $\dfrac{dy}{dx} = \dfrac{\left(\dfrac{dy}{dt}\right)}{\left(\dfrac{dx}{dt}\right)} = \dfrac{3t^2 - 3}{2t} \underset{\text{Make separate fractions so that } \frac{d}{dt}\left(\frac{dy}{dx}\right) \text{ is easy to find.}}{=} \dfrac{3}{2}\left(t - \dfrac{1}{t}\right)$

$\dfrac{d^2y}{dx^2} = \dfrac{\dfrac{d}{dt}\left(\dfrac{dy}{dx}\right)}{\left(\dfrac{dx}{dt}\right)} = \dfrac{\dfrac{d}{dt}\left(\dfrac{3}{2}\left(t - \dfrac{1}{t}\right)\right)}{\left(\dfrac{d(t^2)}{dt}\right)} \underset{\frac{d\left(\frac{1}{t}\right)}{dt} = \frac{d(t^{-1})}{dt} = -\frac{1}{t^2}}{=} \dfrac{\dfrac{3}{2}\left(1 + \dfrac{1}{t^2}\right)}{2t} = \dfrac{3}{4t}\left(1 + \dfrac{1}{t^2}\right) \underset{\text{You may prefer ...}}{=} \dfrac{3(t^2 + 1)}{4t^3}$

(b) $\dfrac{dw}{dz} = \dfrac{\left(\dfrac{dw}{ds}\right)}{\left(\dfrac{dz}{ds}\right)} = \dfrac{-\sin(s)}{\cos(s)} = -\tan(s)$

$\therefore \dfrac{d^2w}{dz^2} = \dfrac{\dfrac{d}{ds}\left(\dfrac{dw}{dz}\right)}{\left(\dfrac{dz}{ds}\right)} = \dfrac{\dfrac{d}{ds}(-\tan(s))}{\left(\dfrac{d(\sin(s))}{ds}\right)} = \dfrac{-\sec^2(s)}{\cos(s)} = -\sec^3(s)$

Polar Coordinates

POSITIVE angles are drawn **COUNTER-CLOCKWISE** from the positive x axis.

NEGATIVE angles are drawn **CLOCKWISE** from the positive x axis.

POSITIVE RADIUS: Given the point P with polar coordinates (r, θ), with $r > 0$, plot P, with the angle θ, r units from the pole (or origin).

NEGATIVE RADIUS: Given the point P with polar coordinates (r, θ), with $r < 0$, plot a point **(not P!)**, with the angle θ, $-r$ units from the pole (or origin). Then **reflect** this point through the pole. **The reflected point is P!**

Example 1) Plot each of the following points which are given in polar coordinates.
(a) $(3, \pi/4)$ (b) $(1, -\pi/3)$ (c) $(2, 0)$ (d) $(2, \pi)$

Solution

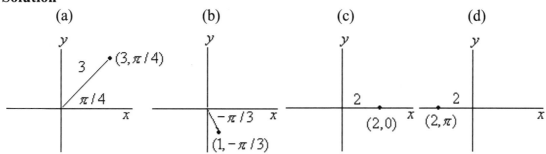

Example 2) Plot each of the following points which are given in polar coordinates.
(a) $(-3, \pi/4)$ (b) $(-1, -\pi/3)$ (c) $(-2, 0)$ (d) $(-2, \pi)$

Solution

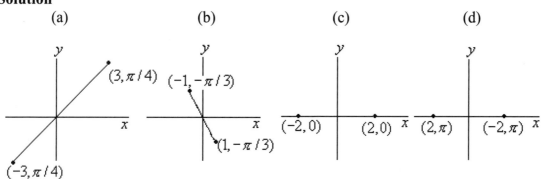

Polar to Rectangular Coordinates; Rectangular to Polar Equations

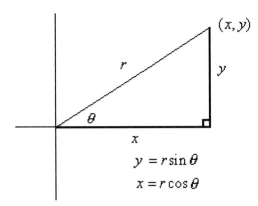

$y = r \sin \theta$
$x = r \cos \theta$

The problem with math teachers when it comes to things like polar and rectangular coordinates is this: how are students to know whether (a, b) is supposed to be in **polar** (a is the radius and b is the angle) coordinates or **rectangular** (a is the horizontal distance and b is the vertical distance) coordinates?

Good answer: teachers would label points something like this: $(a, b)_{PC}$ and $(a, b)_{RC}$!
"Get Real" answer: you are either told explicitly or you pick it up from the context. Sometimes life is cruel and teachers can be (though this is rare) inconsiderate!

Example 1) Convert each of the following points from polar to rectangular coordinates:
(a) $(3, \pi/4)$ (b) $(2, -\pi/3)$ (c) $(-2, 0)$ (d) $(2, \pi/2)$ (e) $(2, -\pi/2)$

Solution (a) $x = 3\cos(\pi/4) = 3\left(\dfrac{1}{\sqrt{2}}\right) = \dfrac{3}{\sqrt{2}}$ and $y = 3\sin(\pi/4) = 3\left(\dfrac{1}{\sqrt{2}}\right) = \dfrac{3}{\sqrt{2}}$

(b) $x = 2\cos(-\pi/3) = 2\left(\dfrac{1}{2}\right) = 1$ and $y = 2\sin(-\pi/2) = 2\left(-\dfrac{\sqrt{3}}{2}\right) = -\sqrt{3}$

(c) $x = -2\cos 0 = 2(1) = -2$ and $y = -2\sin 0 = 0$
(d) $x = 2\cos(\pi/2) = 0$ and $y = 2\sin(\pi/2) = 2$
(e) $x = 2\cos(-\pi/2) = 0$ and $y = 2\sin(-\pi/2) = -2$

Example 2) Convert these rectangular equations to polar equations:
(a) $y = x^2$ (b) $x^2 + y^2 = 16$

Solution (a) $r\sin\theta = r^2 \cos^2\theta$ $\therefore$ $r = 0$ or $r = \dfrac{\sin\theta}{\cos^2\theta} = \tan\theta \sec\theta$.

Note that when $\theta = 0$ in the second equation, $r = 0$. So, we only need $r = \tan\theta \sec\theta$.
(b) $r^2 \cos^2\theta + r^2 \sin^2\theta = 16$ and factoring out r^2, we have $r^2(\cos^2\theta + \sin^2\theta) = 16$. Everybody knows that $\cos^2\theta + \sin^2\theta = 1$ and so the polar equation is $r^2 = 16$. Therefore, $r = 4$ or $r = -4$. However, these are two polar equations for the same curve, the circle around the origin of radius 4. So $r = 4$ will do!

Two For You – Derivatives from Parametric Equations

1)(a) Find $\dfrac{dy}{dx}$ if $x = \ln(t)$ and $y = \tan(t) - t^2$.   (b) Now find $\dfrac{dx}{dy}$ in one step.

2)(a) Find $\dfrac{dy}{dx}$ if $x = \cos^3(\theta)$ and $y = \sin^3(\theta)$.   (b) Now find $\dfrac{dx}{dy}$ in one step.

Answers

1)(a) $\dfrac{dy}{dx} = \dfrac{\sec^2(t) - 2t}{\left(\dfrac{1}{t}\right)} = t\sec^2(t) - 2t^2$ (b) $\dfrac{dx}{dy} = \dfrac{1}{\left(\dfrac{dy}{dx}\right)} = \dfrac{1}{t\sec^2(t) - 2t^2}$

2)(a) $\dfrac{dy}{dx} = \dfrac{\sin^2(\theta)\cos(\theta)}{-\cos^2(\theta)\sin(\theta)} = -\tan(\theta)$ (b) $\dfrac{dx}{dy} = \dfrac{1}{\left(\dfrac{dy}{dx}\right)} = -\cot(\theta)$

Two For You – Higher Derivatives from Parametric Equations

1) Find $\dfrac{d^2y}{dx^2}$ if $x = \ln(t)$ and $y = \tan(t) - t^2$.   Hint: $\dfrac{dy}{dx} = t\sec^2(t) - 2t^2$

2) Find $\dfrac{d^2y}{dx^2}$ if $x = \cos^3(\theta)$ and $y = \sin^3(\theta)$.   Hint: $\dfrac{dy}{dx} = -\tan(\theta)$

Answers

1) $\dfrac{d^2y}{dx^2} = \dfrac{\sec^2(t) + 2t\sec^2(t)\tan(t) - 4t}{\left(\dfrac{1}{t}\right)} = t\sec^2(t) + 2t^2\sec^2(t)\tan(t) - 4t^2$

2) $\dfrac{d^2y}{dx^2} = \dfrac{-\sec^2(\theta)}{-3\cos^2(\theta)\sin(\theta)} = \dfrac{\sec^4(\theta)\csc(\theta)}{3}$

Two For You – Polar Coordinates

1) Plot the points with polar coordinates $(2, 30°)$ and $(2, -30°)$.
2) Plot the points with polar coordinates $(-2, 30°)$ and $(-2, -30°)$.

Answers

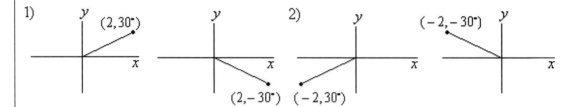

Two For You – Polar to Rectangular Coordinates
Rectangular to Polar Equations

1) Convert these polar coordinates to rectangular: (a) $(5, 5\pi/4)$ (b) $(-2, \pi)$
2) Find a polar equation from the rectangular equation $y = x$.

Answers 1)(a) $\left(-\dfrac{5}{\sqrt{2}}, -\dfrac{5}{\sqrt{2}}\right)$ (b) $(2, 0)$ 2) $\tan\theta = 1$

Rectangular to Polar Coordinates; Polar to Rectangular Equations

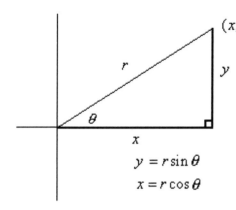

$r^2 = x^2 + y^2$ and so $r = \pm\sqrt{x^2 + y^2}$ $\tan\theta = \dfrac{y}{x}$

Use your calculator or math software to solve $\tan\theta = $ constant, if it isn't an "easy" ratio.

Set your calculator to radians for now : constant 2nd function tan =

$y = r\sin\theta$
$x = r\cos\theta$

Be careful! Your calculator will only give you an answer for θ between $-\pi/2$ and $\pi/2$.

Strategy: (x, y) is...

...in one of the four quadrants
   OR
...on the positive y axis ($x = 0$, $y > 0$)
   OR
...on the negative y axis ($x = 0$, $y < 0$).

The picture at the right shows how to choose r and θ in each case.

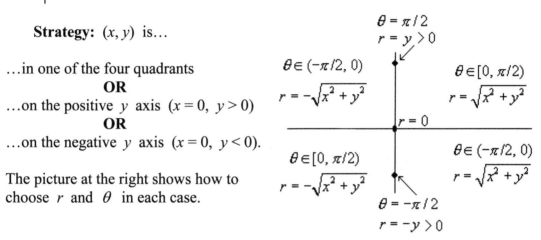

Example 1) Convert each of the points from rectangular to polar coordinates:

(a) $(2, 2)$ (b) $(-1, \sqrt{3})$ (c) $(0, -5)$

Solution (a) $\tan\theta = \dfrac{2}{2} = 1$ and the point is in the first quadrant.

$\therefore$ $\theta = \pi/4$ and $r = +\sqrt{2^2 + 2^2} = \sqrt{8} = 2\sqrt{2}$

(b) $\tan\theta = -\sqrt{3}$ and the point is in the second quadrant.

$\therefore$ $\theta = -\pi/3$ and $r = -\sqrt{(-1)^2 + \sqrt{3}^2} = -\sqrt{4} = -2$

(c) $\tan\theta$ is undefined. Since $y < 0$, we can choose $\theta = -\pi/2$ and $r = 5$.

Example 2) Convert $r\sin\theta = 5$ to an equation in rectangular coordinates.

Solution $\pm\sqrt{x^2 + y^2}\left(\dfrac{y}{\pm\sqrt{x^2 + y^2}}\right) = 5$ which becomes simply $y = 5$.

(Very) Basic Vectors

Let's review the basics of vectors. Vectors are usually denoted with an italicized letter capped by an arrow, for example, $\vec{v}$, or by a bold letter, such as $\boldsymbol{v}$, or both: $\vec{\boldsymbol{v}}$. The length (or magnitude) is denoted $\|\vec{v}\|$.

$\mathbb{R}^2$	$\mathbb{R}^3$
Let $\vec{v}=(a,b)$, $\vec{w}=(c,d)$, and $k \in \mathbb{R}$. $\vec{v} \pm \vec{w} = (a \pm c, b \pm d)$ $k\vec{v} = (ka, kb)$ $\|\vec{v}\| = \sqrt{a^2+b^2}$	Let $\vec{v}=(a,b,c)$, $\vec{w}=(d,e,f)$, and $k \in \mathbb{R}$. $\vec{v} \pm \vec{w} = (a \pm d, b \pm e, c \pm f)$ $k\vec{v} = (ka, kb, kc)$ $\|\vec{v}\| = \sqrt{a^2+b^2+c^2}$

Example 1) Let the vector $\vec{v} = (1,2)$.

(a) Illustrate $\vec{v}$ as a directed line segment beginning at (i) $(0,0)$ (ii) $(-3,-3)$.

(b) Find and illustrate $\vec{w} = -2\vec{v}$ starting at $(0,2)$ using your picture in (a).

(c) Find the length of (i) $\vec{v}$ and (ii) $\vec{w}$.

Solution (a) and (b) $\vec{w} = -2\vec{v} = -2(1,2) = (-2,-4)$

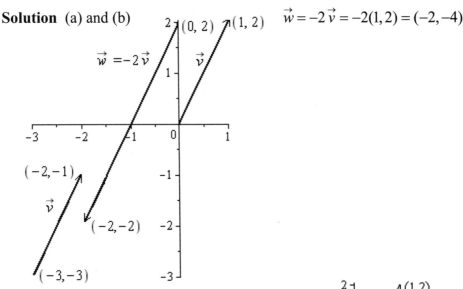

(c)(i) $\|\vec{v}\| = \sqrt{1^2 + 2^2} = \sqrt{1+4} = \sqrt{5}$.

Note that $\vec{v}$ is the hypotenuse of this right triangle.

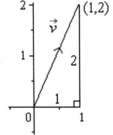

Is $\|\vec{w}\| = 2\|\vec{v}\|$ a surprise? NO! However, notice that the "$-$" goes away!

(c)(ii) $\|\vec{w}\| = \|-2\vec{v}\| = \|(-2,-4)\| = \sqrt{(-2)^2 + (-4)^2} = \sqrt{4+16} = \sqrt{20} = 2\sqrt{5} = 2\|\vec{v}\|$.

The Dot or Scalar or Inner Product of Two Vectors

The dot product is also called the "**scalar product**" because it is a form of vector multiplication which yields a scalar, **not a vector!** Why is it called the "**inner product**" as well? See ** on the last line of the **Definition** below.

Definition The dot or scalar product of the two vectors $\vec{a}$ and $\vec{b}$ is $\vec{a} \cdot \vec{b} = \|\vec{a}\| \|\vec{b}\| \cos\theta$, where θ is the angle between $\vec{a}$ and $\vec{b}$, $0 \leq \theta \leq \pi$. If, as in the diagram, $\vec{a} = (a_1, a_2)$ and $\vec{b} = (b_1, b_2)$, it turns out that
$\vec{a} \cdot \vec{b} = \|\vec{a}\| \|\vec{b}\| \cos\theta = a_1 b_1 + a_2 b_2$.
If we are dealing with more dimensions, $\vec{a} = (a_1, a_2, ..., a_n)$ and $\vec{b} = (b_1, b_2, ..., b_n)$, then $\vec{a} \cdot \vec{b} = a_1 b_1 + a_2 b_2 + ... + a_n b_n$. **

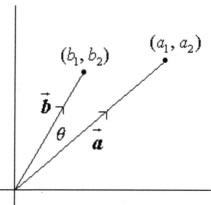

Example 1) Given $\vec{a} = (3, 4, 0)$ and $\vec{b} = (-1, 2, \sqrt{11})$, find
(a) $\vec{a} \cdot \vec{b}$ (b) $\|\vec{a}\|$ and $\|\vec{b}\|$ (c) θ, the angle between $\vec{a}$ and $\vec{b}$.

Solution (a) $\vec{a} \cdot \vec{b} = (3, 4, 0) \cdot (-1, 2, \sqrt{11}) = 3 \times (-1) + 4 \times 2 + 0 \times \sqrt{11} = 5$
(b) $\|\vec{a}\| = \sqrt{3^2 + 4^2 + 0^2} = \sqrt{25} = 5$ $\quad \|\vec{b}\| = \sqrt{(-1)^2 + 2^2 + \sqrt{11}^2} = \sqrt{16} = 4$
(c) From $\vec{a} \cdot \vec{b} = \|\vec{a}\| \|\vec{b}\| \cos\theta = a_1 b_1 + a_2 b_2 + a_3 b_3$
$\cos\theta = \dfrac{a_1 b_1 + a_2 b_2 + a_3 b_3}{\|\vec{a}\| \|\vec{b}\|} = \dfrac{5}{5 \times 4} = \dfrac{1}{4} = 0.25$

| Put your calculator in RADIAN mode! 0.25 SecondFunction cos= | | Put your calculator in DEGREE mode! 0.25 SecondFunction cos= |

$\therefore \theta \doteq 1.32$ radians $\quad$ or $\quad \therefore \theta \doteq 75.5°$

Example 2) What can you conclude about θ or $\vec{a}$ or $\vec{b}$ if $\vec{a} \cdot \vec{b}$ is
(a) 0? (b) $\|\vec{a}\| \|\vec{b}\|$?

Solution (a) If $\vec{a} \cdot \vec{b} = \|\vec{a}\| \|\vec{b}\| \cos\theta = 0$, then either $\vec{a} = \vec{0}$ or $\vec{b} = \vec{0}$ or $\theta = \pi/2$.
(b) If $\vec{a} \cdot \vec{b} = \|\vec{a}\| \|\vec{b}\| \cos\theta = \|\vec{a}\| \|\vec{b}\|$, then $\cos\theta = 1$ and so $\theta = 0$.

The Projection of One Vector on Another

As I write this in July, it is $40°$ C. A little warm. I am remembering temperatures of $-15°$ C. COOL! I prefer cool. You can dress for cool. I am also remembering pushing my car out of a snow bank. Look at the picture below. I would get the best results if I were pushing directly into the back of the car, parallel to the ground. Unfortunately, for traction's sake, I had to dig in my heels and in effect push upwards at the angle θ. We measure the effect of the magnitude of the vector $\vec{u}$ on the vector $\vec{v}$ by dropping a perpendicular from the end of $\vec{u}$ onto $\vec{v}$. The effect of my efforts is measured by AD, called the component of $\vec{u}$ on $\vec{v}$, denoted $\text{comp}_{\vec{v}}\,\vec{u}$. The vector in the direction of $\vec{v}$ with this magnitude is the vector projection of $\vec{u}$ on $\vec{v}$, denoted $\text{proj}_{\vec{v}}\,\vec{u}$.

$$\frac{AD}{\|\vec{u}\|} = \cos(\theta) \Rightarrow \text{comp}_{\vec{v}}\,\vec{u} = AD = \|\vec{u}\|\cos(\theta) \overset{\text{Multiply top and bottom by } \|\vec{v}\|}{=} \frac{\|\vec{u}\|\|\vec{v}\|\cos(\theta)}{\|\vec{v}\|} = \frac{\vec{u}\cdot\vec{v}}{\|\vec{v}\|}$$

$$\text{proj}_{\vec{v}}\,\vec{u} = (\text{comp}_{\vec{v}}\,\vec{u})(\text{a unit vector in the direction of } \vec{v}) = \frac{\vec{u}\cdot\vec{v}}{\|\vec{v}\|}\frac{\vec{v}}{\|\vec{v}\|} = \left(\frac{\vec{u}\cdot\vec{v}}{\vec{v}\cdot\vec{v}}\right)\vec{v}$$

Example 1) Given $\vec{u} = (3, 4, 0)$ and $\vec{v} = \left(-1, 2, \sqrt{11}\right)$, find

(a) $\text{comp}_{\vec{v}}\,\vec{u}$ (b) $\text{proj}_{\vec{v}}\,\vec{u}$ (c) $\text{comp}_{\vec{u}}\,\vec{v}$ (d) $\text{proj}_{\vec{u}}\,\vec{v}$

Solution (a) $\text{comp}_{\vec{v}}\,\vec{u} = \dfrac{(3, 4, 0)\cdot\left(-1, 2, \sqrt{11}\right)}{\|\left(-1, 2, \sqrt{11}\right)\|} = \dfrac{3\times(-1) + 4\times 2 + 0\times\sqrt{11}}{\sqrt{1+4+11}} = \dfrac{5}{4}$

(b) $\text{proj}_{\vec{v}}\,\vec{u} = \dfrac{(3, 4, 0)\cdot\left(-1, 2, \sqrt{11}\right)}{\left(-1, 2, \sqrt{11}\right)\cdot\left(-1, 2, \sqrt{11}\right)}\left(-1, 2, \sqrt{11}\right) = \dfrac{5}{16}\left(-1, 2, \sqrt{11}\right) = \left(-\dfrac{5}{16}, \dfrac{5}{8}, \dfrac{5\sqrt{11}}{16}\right)$

(c) $\text{comp}_{\vec{u}}\,\vec{v} = \dfrac{\left(-1, 2, \sqrt{11}\right)\cdot(3, 4, 0)}{\|(3,4,0)\|} = \dfrac{3\times(-1) + 4\times 2 + 0\times\sqrt{11}}{\sqrt{9+16+0}} = \dfrac{5}{5} = 1$

(d) $\text{proj}_{\vec{u}}\,\vec{v} = \dfrac{\left(-1, 2, \sqrt{11}\right)\cdot(3, 4, 0)}{(3,4,0)\cdot(3,4,0)}(3,4,0) = \dfrac{5}{25}(3,4,0) = \left(\dfrac{3}{5}, \dfrac{4}{5}, 0\right)$

Two For You – Rectangular to Polar Coordinates
Polar to Rectangular Equations

1) Convert rectangular coordinates $(-3,-3)$ to polar coordinates.

2) Find a rectangular equation corresponding to $r = \sin\theta$.

Answers 1) $\theta = \pi/4$ and $r = -3\sqrt{2}$ 2) $x^2 + y^2 = y$ or $x^2 + \left(y - \frac{1}{2}\right)^2 = \frac{1}{4}$

Two For You – (Very) Basic Vectors

1) Let the vector $\vec{v} = (-1, 2)$.

(a) Illustrate $\vec{v}$ as a directed line segment beginning at the **point** (i) $(0,0)$ (ii) $(2,1)$.

(b) Find and illustrate $\vec{w} = -2\vec{v}$ starting at $(0,0)$ using your picture in (a).

(c) Find the length of (i) $\vec{v}$ and (ii) $\vec{w}$.

2) Let $\vec{v} = (1, 2, 3)$ and $\vec{w} = (-2, 1, 3)$.

(a) Find $\vec{v} + \vec{w}$.

(b) Find the lengths of $\vec{v}$, $\vec{w}$ and $\vec{v} + \vec{w}$.

(c) Note that $\|\vec{v} + \vec{w}\| < \|\vec{v}\| + \|\vec{w}\|$. The three vectors $\vec{v}$, $\vec{w}$ and $\vec{v} + \vec{w}$ can be positioned to form the sides of a triangle. The inequality is an example of the **Side Inequality Theorem: The sum of the lengths of two sides of a triangle is always…** (finish this statement.)

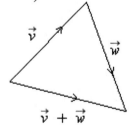

Answers 1)(a) and (b) 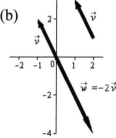 (c) $\|\vec{v}\| = \sqrt{5}$ $\|\vec{w}\| = 2\sqrt{5}$

2)(a) $(-1, 3, 6)$ (b) $\|\vec{v}\| = \sqrt{14}$ $\|\vec{w}\| = \sqrt{14}$ $\|\vec{v} + \vec{w}\| = \sqrt{46}$

(c) …greater than the length of the third side.

Two For You – Dot or Scalar or Inner Product of Two Vectors

1) Find the angle between $\vec{a}$ and $\vec{b}$ if $\vec{a} = (1, -1)$ and $\vec{b} = (-1, 2)$.

2) What can you conclude about θ or $\vec{a}$ or $\vec{b}$ if the value of $\vec{a} \cdot \vec{b}$ is $-\|\vec{a}\|\|\vec{b}\|$?

Answers 1) $\theta \doteq 2.8$ radians or $\theta \doteq 161.6°$ 2) $\theta = \pi$ or $\vec{a} = \vec{0}$ or $\vec{b} = \vec{0}$

Two For You – The Projection of One Vector on Another

1) Given $\vec{u} = (1, 2)$ and $\vec{v} = (3, 4)$, find

(a) $\text{comp}_{\vec{v}} \vec{u}$ (b) $\text{proj}_{\vec{v}} \vec{u}$ (c) $\text{comp}_{\vec{u}} \vec{v}$ (d) $\text{proj}_{\vec{u}} \vec{v}$

2) Assume $\vec{u}$ and $\vec{v}$ non-zero vectors.

(a) When is $\text{comp}_{\vec{v}} \vec{u} = \|\vec{u}\|$?   (b) When is $\text{comp}_{\vec{v}} \vec{u} = 0$?   (c) When is $\text{comp}_{\vec{v}} \vec{u} = \vec{0}$?

Answers 1)(a) $\dfrac{11}{5}$ (b) $\left(\dfrac{33}{25}, \dfrac{44}{25}\right)$ (c) $\dfrac{11}{\sqrt{5}}$ (d) $\left(\dfrac{11}{5}, \dfrac{22}{5}\right)$

2(a) $\theta = 0°$ (b) $\theta = 90°$ (c) Never: $\text{comp}_{\vec{v}} \vec{u}$ is a scalar, not a vector!

The Vector or Cross Product of Two Vectors

The cross product is also called the "**vector product**" because it is a type of vector multiplication which, unlike the dot product, **does yield a vector!** In fact, the cross product of $\vec{a}$ and $\vec{b}$ (which are vectors in $\mathbb{R}^3$) is a vector **perpendicular to the plane** (when $\vec{a}$ and $\vec{b}$ are linearly independent) spanned by $\vec{a}$ and $\vec{b}$!

Definition The cross or vector product of two vectors $\vec{a}$ and $\vec{b}$ is a vector with magnitude $\|\vec{a} \times \vec{b}\| = \|\vec{a}\| \|\vec{b}\| \sin\theta$, where θ is the angle between $\vec{a}$ and $\vec{b}$, $0 \leq \theta \leq \pi$.

This range for θ **guarantees** $\|\vec{a} \times \vec{b}\| = \|\vec{a}\| \|\vec{b}\| \sin\theta \geq 0$!

The direction of $\vec{a} \times \vec{b}$ is given by the **Right Hand Rule**: Let the fingers of your right hand curl in the direction of rotation from $\vec{a}$ to $\vec{b}$. Your thumb points in the direction of $\vec{a} \times \vec{b}$.

If $\vec{a} = (a_1, a_2, a_3)$ and $\vec{b} = (b_1, b_2, b_3)$, it turns out that

$$\vec{a} \times \vec{b} = (a_2 b_3 - a_3 b_2, \, a_3 b_1 - a_1 b_3, \, a_1 b_2 - a_2 b_1) = \begin{vmatrix} \vec{i} & \vec{j} & \vec{k} \\ a_1 & a_2 & a_3 \\ b_1 & b_2 & b_3 \end{vmatrix}.$$

(This is for those of you who are comfortable with the **DETERMINANT** of a 3×3 matrix.)

> Here is an easy pattern for $\vec{a} \times \vec{b}$!
> The **1st** coordinate uses **2, 3**; the **2nd** uses **3, 1**; the **3rd** uses **1, 2**.

Example 1) Given $\vec{a} = (1, 2, 3)$ and $\vec{b} = (4, 5, 6)$, find $\vec{a} \times \vec{b}$. Which quadrant is θ in?

Solution $\vec{a} \times \vec{b} = (1, 2, 3) \times (4, 5, 6) = (2 \times 6 - 5 \times 3, \, 3 \times 4 - 6 \times 1, \, 1 \times 5 - 4 \times 2) = (-3, 6, -3)$

Note that $\vec{a} \cdot \vec{b} = 4 + 10 + 18 = 32 > 0$.

$\therefore \cos\theta = \dfrac{\vec{a} \cdot \vec{b}}{\|\vec{a}\| \|\vec{b}\|} > 0$ and so $\theta \in (0, \pi/2)$. (We used this formula on the page 233!)

Example 2) Let $\vec{a} = (a_1, a_2, a_3) \neq \vec{0}$, $\vec{b} = (b_1, b_2, b_3) \neq \vec{0}$, and $\vec{a} \not\parallel \vec{b}$, so $\vec{a} \times \vec{b} \neq \vec{0}$.
Prove $\vec{a} \times \vec{b} \perp \vec{a}$. (We define $\vec{0}$ to be orthogonal to all vectors.)

Solution $(\vec{a} \times \vec{b}) \cdot \vec{a} = (a_2 b_3 - a_3 b_2, \, a_3 b_1 - a_1 b_3, \, a_1 b_2 - a_2 b_1) \cdot (a_1, a_2, a_3)$

$= a_1 a_2 b_3 - a_1 a_3 b_2 + a_2 a_3 b_1 - a_2 a_1 b_3 + a_3 a_1 b_2 - a_3 a_2 b_1 = 0$. (The terms cancel out in pairs!) If θ is the angle between $\vec{a} \times \vec{b}$ and $\vec{a}$, then $(\vec{a} \times \vec{b}) \cdot \vec{a} = \|\vec{a} \times \vec{b}\| \|\vec{a}\| \cos\theta$ $\therefore \cos\theta = 0$ and so $\theta = \dfrac{\pi}{2}$.

Vector Equation of a Line: $\overrightarrow{OP} = \overrightarrow{OP_0} + t\vec{v}$

To find the vector equation of a line, we need:
(i) a point P_0 on the line and

(ii) a vector $\vec{v}$ parallel to the line.
Then, if P is any point on the line, we can write

$\overrightarrow{OP} = \overrightarrow{OP_0} + t\vec{v}$. ($\overrightarrow{OP_0}$ gets us **onto the line** and

then $t\vec{v}$ lets us **travel along the line**!)

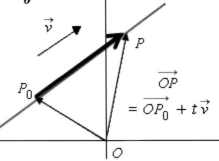

Scalar equ'n: $Ax + By = C$ gives the set of all points (x, y) on the line.

Vector equ'n: $\overrightarrow{OP} = \overrightarrow{OP_0} + t\vec{v}$ gives the set of all vectors from $(0,0)$ to the line.

Example 1) Find (a) vector (b) parametric and (c) symmetric equations of the line through $P_0(1, 2)$ and parallel to $\vec{v} = (2, -3)$.

Solution (a) **Vector Equation**: $\overrightarrow{OP}$ [$\overrightarrow{OP} = \overrightarrow{OP_0} + t\vec{v}$] $= (1, 2) + t(2, -3)$, or $(x, y) = (1, 2) + t(2, -3)$

(b) **Parametric Equations**: x [Equate the first coordinates in the vector equation.] $= 1 + 2t$, y [Equate the second coordinates in the vector equation.] $= 2 - 3t$

(c) **Symmetric Equations**: t [Solve for t in the parametric equations.] $= \dfrac{x-1}{2} = \dfrac{y-2}{-3}$

Example 2) Find (a) vector (b) parametric and (c) symmetric equations of the line through $P_0(1, 2, -1)$ and parallel to $\vec{v} = (2, -3, 4)$.

Solution (a) **Vector**: $\overrightarrow{OP} = (1, 2, -1) + t(2, -3, 4)$ or $(x, y, z) = (1, 2, -1) + t(2, -3, 4)$.

(b) **Parametric**: $x = 1 + 2t$, $y = 2 - 3t$, $z = -1 + 4t$

(c) **Symmetric**: $t = \dfrac{x-1}{2} = \dfrac{y-2}{-3} = \dfrac{z+1}{4}$

Example 3) Find (a) vector (b) parametric and (c) symmetric equations of the line through $P_0(1, 2, -1)$ and parallel to $\vec{v} = (2, 0, 0)$.

Solution (a) **Vector**: $\overrightarrow{OP} = (1, 2, -1) + t(2, 0, 0)$, or $(x, y, z) = (1, 2, -1) + t(2, 0, 0)$.

(b) **Parametric**: $x = 1 + 2t$, $y = 2$, $z = -1$

(c) **Symmetric**: $t = \dfrac{x-1}{2}$, $y = 2$, $z = -1$

Note: Here, y and z are constant and so there is only the one symmetric equation. This line is parallel to the x axis, through the point $(1, 2, -1)$. For comparison, the x axis itself, in $\mathbb{R}^3$ has symmetric equation $t = x$ with $y = 0$, $z = 0$.

Vector Equation of a Plane: $\overrightarrow{OP} = \overrightarrow{OP_0} + s\vec{v} + t\vec{w}$

To find the vector equation of a plane, we need:

(i) a point P_0 on the plane and
(ii) two **NON-PARALLEL** direction vectors
$\vec{v}$ and $\vec{w}$ for the plane.
Then, if P is any point on the plane, we can write
$\overrightarrow{OP} = \overrightarrow{OP_0} + s\vec{v} + t\vec{w}$. ($\overrightarrow{OP_0}$ gets us **onto the plane**
and then $s\vec{v} + t\vec{w}$ lets us **travel along the plane to
the required point** P!)

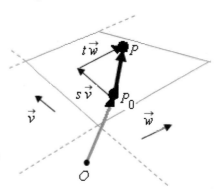

Example 1) Find (a) vector and (b) parametric equations of the plane through $P_0(1,2,1)$ and parallel to vectors $\vec{v} = (2,-3,1)$ and $\vec{w} = (0,2,3)$.

Solution Note $\vec{v}$ and $\vec{w}$ are independent (which, for two vectors, means non-parallel!)

> You should take the time NOW to show that $\vec{v}$ and $\vec{w}$ are independent/non-parallel, that is, show $a\vec{v} + b\vec{w} = (0,0,0) \Rightarrow a = b = 0$.
> In fact, you are showing that neither $\vec{v}$ nor $\vec{w}$ is a multiple of the other.

(a) **Vector Equation**:

$\overrightarrow{OP}$ = $\boxed{\overrightarrow{OP}=\overrightarrow{OP_0}+s\vec{v}+t\vec{w}}$ $(1,2,1) + s(2,-3,1) + t(0,2,3)$ or $(x,y,z) = (1,2,1) + s(2,-3,1) + t(0,2,3)$

> For x, equate the first coordinates in the vector equation; for y, equate the second coordinates; for z, the third

(b) **Parametric Equations:** $x = 1 + 2s$, $y = 2 - 3s + 2t$, $z = 1 + s + 3t$

(Just in case you are wondering, we don't have symmetric equations of a plane!)

Example 2) Find (a) vector and (b) parametric equations of the plane through $P_0(0,0,0)$ and parallel to the plane given by $\overrightarrow{OP} = (1,2,1) + s(1,0,0) + t(0,1,0)$

Solution For two independent direction vectors, we can use
$\vec{v} = (1,0,0)$ and $\vec{w} = (0,1,0)$.

(a) **Vector Equation**: $\overrightarrow{OP} = (0,0,0) + s(1,0,0) + t(0,1,0) = s(1,0,0) + t(0,1,0)$
(b) **Parametric Equations:** $x = s$, $y = t$, $z = 0$ **Note:** This is the xy plane in $\mathbb{R}^3$.

The Scalar Equation of a Plane: $Ax+By+Cz=D$

A mathematics philosophical digression: When a mathematical explorer is first venturing into new territory, she has to define what she deems to be critical concepts and then conjecture and prove theorems that connect those concepts. Unlike the texts that come later, this is rarely a smooth, linear process. But when brilliant insight leads to just the right concepts, intellectual power and beauty emerge. Someone defined the dot product and showed that the dot product of perpendicular vectors is 0. From that simple yet profound observation…

Let $\vec{n}=(A,B,C)$ be perpendicular to a plane (that is, $\vec{n}$ is **normal** to the plane) and $P_0(x_0,y_0,z_0)$ be a given point on the plane. If $P(x,y,z)$ is any point on the plane,

$\vec{n} \perp \overrightarrow{P_0P}$ and so $\vec{n} \cdot \overrightarrow{PP_0} = 0$.

So $(A,B,C)\cdot(x-x_0, y-y_0, z-z_0) = 0$.
This leads to $Ax+By+Cz=D$,
where $D = Ax_0 + By_0 + Cz_0$.

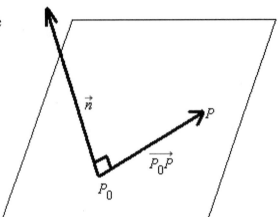

…we get the equation of a plane. Simple, yet powerful.

Using $\vec{n}=(A,B)$ and point $P_0(x_0,y_0)$, the same process shows that the equation of a line is $Ax+By=C$.

Example 1) Find the scalar equation of the plane through $P_0(1,2,1)$ with normal $\vec{n}=(2,-3,1)$.

Solution If $P(x,y,z)$ is any point on the plane, then $\vec{n} \perp \overrightarrow{P_0P}$ and so

$\vec{n} \cdot \overrightarrow{PP_0} = 0$, ie., $(2,-3,1)\cdot(x-1, y-2, z-1) = 0$

$\therefore 2x-2-3y+6+z-1=0$ and so $2x-3y+z=-3$

Example 2) What is the best way to find the scalar equation of a plane if you are given three **non-collinear** points P, Q, and R?

Solution Because the points are non-collinear, $\overrightarrow{PQ}$ and $\overrightarrow{PR}$ are non-parallel direction vectors for the plane. Therefore, $\vec{n} = \overrightarrow{PQ} \times \overrightarrow{PR}$ is a normal to the plane. Now using normal $\vec{n}$ and any one of the points P, Q, and R, proceed as in Example 1.

Two For You – The Vector or Cross Product of Two Vectors

1) Find $\vec{a} \times \vec{b}$ if $\vec{a} = (1, 2, 3)$ and $\vec{b} = (1, 2, 3)$.

2) What can you conclude about $\vec{a} \times \vec{b}$, if anything, if $\vec{a}$ is parallel to $\vec{b}$?

Answers 1) $\vec{a} \times \vec{b} = \vec{0}$ 2) $\vec{a} \times \vec{b} = \vec{0}$

Two For You – Vector Equation of a Line: $\overrightarrow{OP} = \overrightarrow{OP_0} + t\vec{v}$

1) Find (a) vector (b) parametric and (c) symmetric equations of the line through $P_0(-1, 3)$ and parallel to $\vec{v} = (2, 1)$.

2) Find (a) vector (b) parametric and (c) symmetric equations of the line through $P_0(1, 2, -1)$ and parallel to the line given by $\overrightarrow{OP} = (3, 5, -3) + t(2, -3, 4)$.

Answers

1) **Vector:** $\overrightarrow{OP} = (-1, 3) + t(2, 1)$
   Parametric: $x = -1 + 2t,\ y = 3 + t$
   Symmetric: $t = \dfrac{x+1}{2} = \dfrac{y-3}{1}$

2) **Vector:** $\overrightarrow{OP} = (1, 2, -1) + t(2, -3, 4)$
   Parametric: $x = 1 + 2t,\ y = 2 - 3t,\ z = -1 + 4t$
   Symmetric: $t = \dfrac{x-1}{2} = \dfrac{y-2}{-3} = \dfrac{z+1}{4}$

Two For You – Vector Equation of a Plane: $\overrightarrow{OP} = \overrightarrow{OP_0} + s\vec{v} + t\vec{w}$

1) Find (a) vector and (b) parametric equations of the plane through $P_0(-2,3,5)$ and parallel to vectors $\vec{v} = (3,2,1)$ and $\vec{w} = (2,5,1)$.

2) Find (a) vector and (b) parametric equations of the plane through $P_0(0,0,1)$ and perpendicular to the planes given by $x+y+z=3$ and $-x+2y-4z=7$.
(Hint: A normal to, for example, the plane given by $x+y+z=3$ is a direction vector for the required plane.)

Answers 1)(a) **vector:** $\overrightarrow{OP} = (-2,3,5) + s(3,2,1) + t(2,5,1)$
(b) **parametric:** $x = -2 + 3s + 2t,\ y = 3 + 2s + 5t,\ z = 5 + s + t$
2)(a) **vector:** $\overrightarrow{OP} = (0,0,1) + s(1,1,1) + t(-1,2,-4)$
(b) **parametric:** $x = s - t,\ y = s + 2t,\ z = 1 + s - 4t$

Two For You - The Scalar Equation of a Plane: $Ax + By + Cz = D$

1) Find the scalar equation of the plane through $(1,2,3)$ and parallel to the plane $-2x + 3z = 7$. (Hint: a normal to this plane is $(-2,0,3)$.)

2) Find the scalar equation of the plane with vector equation $\overrightarrow{OP} = (4,1,3) + s(1,2,3) + t(1,-2,-4)$. (Hint: take the **cross product** of the two direction vectors to obtain a **normal** to the plane.)

Answers 1) $-2x + 3z = 7$ 2) $-2x + 7y - 4z = -13$

Intersection of Two Lines in $\mathbb{R}^3$: Parallel/Coincident Case

With two lines in $\mathbb{R}^3$, there are four possible scenarios. The lines can be:
(1) coincident   (2) parallel   (3) intersect in exactly one point   (4) "skew".
Here we will deal with (1) and (2).

Critical Note: When the direction vectors from two vector equations of lines are multiples of one another, the lines **must** have…

...**NO** points in common (**parallel** lines) or **ALL** points in common (**coincident** lines).

So, if we have two such direction vectors, take **ANY** point on one of the lines. If the point is on the other line, the lines are coincident. If not, the lines are parallel.

Another Critical Note: In the first **Critical Note**, be careful to distinguish between lines and vectors. Vectors have direction. Lines don't.

Geometry	Algebraic Example	What to do
(1) The two lines are coincident.	L1: $\overrightarrow{OP} = (1,1,1) + s(1,2,3)$ L2: $\overrightarrow{OP} = (-1,-3,-5) + t(-2,-4,-6)$	Note that the two direction vectors are **multiples of one one another** and then show there **is** a common point.
(2) The two lines are parallel.	L1: $\overrightarrow{OP} = (1,1,1) + s(1,2,3)$ L2: $\overrightarrow{OP} = (-1,-3,-4) + t(-2,-4,-6)$	Note that the two direction vectors are **multiples of one another** and then show there **is no** common point.

(1) L1 has direction vector $(1,2,3)$ and L2 has direction vector $(-2,-4,-6)$. These are clearly multiples of one another, so L1 and L2 are either parallel or coincident. Letting $s = 0$ shows $(1,1,1)$ is a point on L1. Now, let's check whether it is also on L2.

$(1,1,1) = (-1,-3,-5) + t(-2,-4,-6)$

$\Leftrightarrow 1 = -1 - 2t \quad\quad 1 = -3 - 4t \quad\quad 1 = -5 - 6t$

$\Leftrightarrow 2 = -2t \quad\quad\quad\; 4 = -4t \quad\quad\quad\; 6 = -6t$

$\Leftrightarrow t = -1 \quad\quad\quad\;\; t = -1 \quad\quad\quad\;\; t = -1$

$\therefore (1,1,1) = (-1,-3,-5) - (-2,-4,-6)$ and so $(1,1,1)$ is on L2.
So this point **is on both lines**. Therefore, the lines are coincident.

(2) The direction vectors are clearly multiples of each other. $s = 0$ shows $(1,1,1)$ is a point on L1. Now, let's check whether it is also on L2.

$(1,1,1) = (-1,-3,-4) + t(-2,-4,-6)$

$\Leftrightarrow 1 = -1 - 2t \quad\quad 1 = -3 - 4t \quad\quad 1 = -4 - 6t$

$\Leftrightarrow 2 = -2t \quad\quad\quad\; 4 = -4t \quad\quad\quad\; 5 = -6t$

$\Leftrightarrow t = -1 \quad\quad\quad\;\; t = -1 \quad\quad\quad\;\; t = -\dfrac{5}{6}$. Since we do not have a common t value, $(1,1,1)$ cannot be on L2. Therefore, L1 is parallel to L2.

Intersection of Two Lines in $\mathbb{R}^3$: Non-Parallel/Non-Coincident Case

With two lines in $\mathbb{R}^3$, there are four possible scenarios. The lines can be (1) coincident or (2) parallel or (3) intersect in exactly one point or (4) "skew". Here we will deal with (3) and (4).

Critical Note: When the direction vectors from two vector equations of lines are **NOT** multiples of one another, the lines **must** intersect in a unique point or not at all (skew).

Geometry	Algebraic Example	What to do
(3) The two lines intersect in a single point.	L1: $\overrightarrow{OP} = (1,1,1) + s(1,2,3)$ L2: $\overrightarrow{OP} = (-2,0,2) + t(1,1,1)$	Note the direction vectors are not multiples of one another. Equate the x, y, and z parametric equations of each line and then **row reduce** to find a **unique** solution for s and t.
(4) The two lines do not intersect ("skew" lines).	L1: $\overrightarrow{OP} = (1,1,1) + s(1,2,3)$ L2: $\overrightarrow{OP} = (-2,0,1) + t(1,1,1)$	Note the direction vectors are not multiples of one another. Equate the x, y, and z parametric equations of each line and then **row reduce** to show the system of equations is inconsistent.

3) The direction vectors clearly are **not** multiples of each other.

Parametric Equations for L1: $x = 1+s,\ y = 1+2s,\ z = 1+3s$

Parametric Equations for L2: $x = -2+t,\ y = 0+t,\ z = 2+t$

$\begin{array}{|l|} \hline x = 1+s = -2+t \\ y = 1+2s = t \\ z = 1+3s = 2+t \\ \hline \end{array}$ Rewrite these three equations as equations in variables s and t and then **row reduce** the new equations. SEE PAGE 47 TO REVIEW ROW REDUCING A SYSTEM OF EQUATIONS! $\Leftrightarrow$ $\begin{array}{|ll|} \hline \text{E1:} & s - t = -3 \\ \text{E2:} & 2s - t = -1 \\ \text{E3:} & 3s - t = 1 \\ \hline \end{array}$ $\begin{array}{l} \text{E4 = E1} \\ \text{E5 = -2E1 + E2} \\ \text{E6 = -3E1 + E3} \end{array}$ $\Leftrightarrow$ $\begin{array}{|ll|} \hline \text{E4:} & s - t = -3 \\ \text{E5:} & t = 5 \\ \text{E6:} & 2t = 10 \\ \hline \end{array}$

So $t = 5$ and $s = -3 + 5 = 2$. Subbing into either set of parametric equations gives intersection point $(3, 5, 7)$.

4) The direction vectors are **not** multiples of each other.

Parametric Equations for L1: $x = 1+s,\ y = 1+2s,\ z = 1+3s$

Parametric Equations for L2: $x = -2+t,\ y = 0+t,\ z = 1+t$

$\begin{array}{|l|} \hline 1+s = -2+t \\ 1+2s = t \\ 1+3s = 1+t \\ \hline \end{array} \Rightarrow \begin{array}{|ll|} \hline \text{E1:} & s-t = -3 \\ \text{E2:} & 2s-t = -1 \\ \text{E3:} & 3s-t = 0 \\ \hline \end{array}$ $\begin{array}{l}\text{E4 = E1}\\\text{E5 = -2E1 + E2}\\\text{E6 = -3E1 + E3}\end{array}$ $\Leftrightarrow \begin{array}{|ll|} \hline \text{E4:} & s-t = -3 \\ \text{E5:} & t = 5 \\ \text{E6:} & 2t = 9 \\ \hline \end{array}$ $\begin{array}{l}\text{E7 = E4}\\\text{E8 = E5}\\\text{E9 = -2E5 + E6}\end{array}$ $\Leftrightarrow \begin{array}{|ll|} \hline \text{E7:} & s-t = -3 \\ \text{E8:} & t = 5 \\ \text{E9:} & 0 = -1 \\ \hline \end{array}$

E9 shows the system has no solution (**inconsistent!**) and so there is no intersection point.

Think of two swimmers where one swims the **length** of a pool doing the **front crawl** while the other swims the **width** of the pool **underwater**. The two trace out (perpendicular) non-intersecting lines, that is, skew lines!

Intersection of Two Planes

Note: For this topic, we will work with **scalar** equations of planes.
When we have two planes in $\mathbb{R}^3$, there are three possibilities:

Remember: Plane $Ax + By + Cz = D$ has normal $\vec{n} = (A, B, C)$.

Geometry	Algebraic Example	Picture
(1) The two planes are coincident, that is, they are the same plane. This only happens if the two equations are **exact multiples of one another!**	E1: $x + y - z = 1$ E2: $2x + 2y - 2z = 2$ Here, $E2 = 2 \times E1$.	
(2) The two planes are parallel. This only happens if the **normals** of the two equations are **exact multiples of one another but the equations are not!**	E1: $x + y - z = 1$ E2: $2x + 2y - 2z = 3$ $\vec{n_1} = (1, 1, -1); \vec{n_2} = (2, 2, -2)$ Here, $\vec{n_2} = 2\vec{n_1}$ but E2 is not a multiple of E1!	
(3) The two planes are not parallel. In this case, they **must** intersect in a straight line. The normals will **not** be multiples of one another.	E1: $x + y - z = 1$ E2: $x + 2y - 2z = 1$	

Example 1) For each of the following pairs of equations of planes, decide whether the planes are coincident, parallel, or intersect in a line. If they intersect, find a set of parametric equations and a vector equation of the line.

(a)	(b)	(c)
E1: $\quad x + 2y + 3z = 4$	E1: $\quad x + 2y + 3z = 4$	E1: $x - 2y + 3z = 4$
E2: $-4x - 8y - 12z = -16$	E2: $-4x - 8y - 12z = 10$	E2: $\quad x - y + 4z = -1$

Solution

(a) Since $E2 = -4E1$, the planes are coincident.

(b) Since $\vec{n_2} = -4\vec{n_1}$ but $E2 \neq -4E1$, the planes are parallel.

(c) Since the normals are not multiples of one another, the planes intersect in a line.

E1: $x - 2y + 3z = 4$
E2: $\quad x - y + 4z = -1$

$\begin{matrix} E3 = E1 \\ E4 = -E1+E2 \end{matrix}$ $\Leftrightarrow$ SEE PAGE 47 TO REVIEW ROW REDUCING A SYSTEM OF EQUATIONS!

E3: $x - 2y + 3z = 4$
E4: $\qquad y + z = -5$

From E4, we have $y = -5 - z$. Substitute this into E1:

$x - 2(-5 - z) + 3z = 4 \Leftrightarrow x + 10 + 5z = 4 \Leftrightarrow x = -6 - 5z \quad$ Let $z = t$.

Parametric Equations : $x = -6 - 5t, \ y = -5 - t, \ z = t$

$\therefore$ We have point $(-6, -5, 0)$ and, from the t coefficients, direction vector $(-5, -1, 1)$.

Vector Equation : $(x, y, z) = (-6, -5, 0) + t(-5, -1, 1)$

Intersection of Three Planes: Parallel/Coincident Case

Note: For this topic, we will work with **scalar** equations of planes.

There are three **crucial** facts from **The Intersection of Two Planes:**

I) Equation E1 is a multiple of equation E2 $\Leftrightarrow$ The two planes are coincident.

II) E1 and E2 aren't multiples but normals $\vec{n_1}$ and $\vec{n_2}$ are $\Leftrightarrow$ The two planes are ∥.

III) Otherwise, the two planes must intersect in a line.

There are **EIGHT** possible scenarios with three planes. Here, we will deal with the five where at least two of the planes are either parallel or coincident. The very good news: you can determine the answer *just by looking* at the equations. It's easy!

Planes E1, E2, E3 (with normals $\vec{n_1}, \vec{n_2}, \vec{n_3}$) ...	...which, row reduced*, become... *See page 47.	Geometrical description	Picture	What are multiples of what?
(1) $x+y+z=1$ $2x+2y+2z=2$ $3x+3y+3z=3$	$x+y+z=1$ $0=0$ $0=0$	3 coincident planes		E1, E2, E3; $\vec{n_1}, \vec{n_2}, \vec{n_3}$
(2) $x+y+z=1$ $2x+2y+2z=2$ $3x+3y+3z=4$	$x+y+z=1$ $0=0$ $0=-1$	2 coincident planes and a ∥ plane		E1, E2; $\vec{n_1}, \vec{n_2}, \vec{n_3}$
(3) $x+y+z=1$ $x+2y+2z=2$ $2x+2y+2z=2$	$x+y+z=1$ $y+z=1$ $0=0$	2 coincident planes intersecting the 3rd in a line**		E1, E3; $\vec{n_1}, \vec{n_3}$
(4) $x+y+z=1$ $x+y+z=2$ $x+y+z=3$	$x+y+z=1$ $0=-1$ $0=-2$	3 ∥ planes		$\vec{n_1}, \vec{n_2}, \vec{n_3}$
(5) $x+y+z=1$ $x+2y+3z=4$ $x+y+z=2$	$x+y+z=1$ $y+2z=3$ $0=-1$	2 ∥ planes, each intersecting the 3rd plane in a line		$\vec{n_1}, \vec{n_3}$

See **Intersection of Two Planes (page 245) for details on finding the line of intersection.

So in all five parallel/coincident cases, you can just eyeball the **original** equations (you don't need to row reduce!) and the answer will be obvious. For example, in (5), the first and third planes are parallel because their normals are multiples but their equations are not. The middle normal is not a multiple of either of the others so the middle plane must intersect the first in a line and the third in a line. Planes one and three never meet so neither do these intersection lines. In fact, we can say more: the intersection lines are parallel. How can you prove this?

Two For You – Two Lines in $\mathbb{R}^3$: Parallel/Coincident

1)(a) Are these two lines coincident, parallel, skew, or do they intersect in a single point? If they intersect, find the intersection point.

L1: $\overrightarrow{OP} = (1,1,1) + s(1,1,1)$

L2: $\overrightarrow{OP} = (-1,-3,-5) + t(4,4,4)$

(b) If the lines were coincident, what value of s shows that $(-1,-3,-5)$ is on L1?

2)(a) Are these two lines coincident, parallel, skew, or do they intersect in a single point? If they intersect, find the intersection point.

L1: $\overrightarrow{OP} = (19,17,15) + s(1,1,1)$

L2: $\overrightarrow{OP} = (-1,-3,-5) + t(4,4,4)$

(b) If the lines are coincident, what value of s shows that $(-1,-3,-5)$ is on L1?

Answers 1)(a) parallel (b) non applicable since $(-1,-3,-5)$ is **not** on L1.
2)(a) coincident (b) $s = -20$

Two For You – Two Lines in $\mathbb{R}^3$: Non-Parallel/Non-Coincident

1)(a) Are these two lines coincident, parallel, skew, or do they intersect in a single point?

L1: $\overrightarrow{OP} = (1,1,1) + s(1,2,3)$

L2: $\overrightarrow{OP} = (8,3,-2) + t(3,2,1)$

(b) If they intersect in a single point, find this point. Give the values of s and t in L1 and L2 respectively which produce this intersection point.

2)(a) Are these two lines coincident, parallel, skew, or do they intersect in a single point? If they intersect, find the intersection point.

L1: $\overrightarrow{OP} = (1,1,1) + s(1,2,3)$

L2: $\overrightarrow{OP} = (-1,-3,5) + t(3,2,1)$

(b) If they intersect in a single point, find this point. Give the values of s and t in L1 and L2 respectively which produce this intersection point.

Answers 1)(a) intersect in a single point (b) $(-1,-3,-5)$, $s = -2$, $t = -3$
2)(a) skew (b) not applicable since the lines do not intersect.

Three For You – Intersection of Two Planes

1) (a) Are these two planes coincident, parallel, or do they intersect in a line?
   E1: $5x+2y-z=1$ E2: $-10x-4y+2z=0$
   (b) If they intersect in a line, find a set of parametric equations for the line.
2) (a) Are these two planes coincident, parallel, or do they intersect in a line?
   E1: $5x+2y-z=1$ E2: $-10x-4y+2z=-2$
   (b) If they intersect in a line, find a set of parametric equations for the line.
3) (a) Are these two planes coincident, parallel, or do they intersect in a line?
   E1: $x-y+4z=1$ E2: $5x+2y-z=-2$
   (b) If they intersect in a line, find a set of parametric equations for the line.

Answers 1)(a) parallel (The normals are multiples but the equations are not.)
(b) not applicable since the planes do not intersect
2)(a) coincident (The equations are multiples of each other.)
(b) not applicable since the planes are coincident
3)(a) The planes intersect in a line.   (b) $x=-t,\ y=-1+3t,\ z=t$

Four For You – Intersection of Three Planes: Parallel/Coincident

Complete the table.

Three Planar Equations	(a) Which equations and which normals are multiples of one another?	(b) Geometric Description
1) E1: $3x+2y+z=1$ E2: $6x+4y+2z=2$ E3: $3x+3y+3z=3$		
2) E1: $3x+2y+z=1$ E2: $6x+4y+2z=5$ E3: $-3x-2y-z=3$		
3) E1: $3x+2y+z=1$ E2: $6x+4y+2z=2$ E3: $9x+6y+3z=5$		
4) E1: $3x+2y+z=1$ E2: $6x+4y+2z=-2$ E3: $3x+3y+3z=3$		

Answer 1)(a) E1, E2; $\vec{n_1}, \vec{n_2}$ (b) 2 coincident planes intersecting the 3rd in a line

2)(a) $\vec{n_1}, \vec{n_2}, \vec{n_3}$ (b) 3 ∥ planes   3)(a) E1, E2; $\vec{n_1}, \vec{n_2}, \vec{n_3}$ (b) 2 coincident planes and a ∥ plane

4)(a) $\vec{n_1}, \vec{n_2}$ (b) 2 ∥ planes, each intersecting the 3rd plane in a line

Intersection of Three Planes: Non-Parallel/Non-Coincident Case

Note: For this topic, we will work with **scalar** equations of planes.

There are three **crucial** facts from **The Intersection of Two Planes**:
I) E1 is a multiple of E2 $\Leftrightarrow$ The two planes are coincident.

II) E1 and E2 aren't multiples but $\vec{n_1}$ and $\vec{n_2}$ are $\Leftrightarrow$ The two planes are parallel.

III) Otherwise, the two planes must intersect in a line.

There are **EIGHT** possible scenarios with three planes. We dealt with the five parallel/coincident cases in the previous section. Here, we will deal with the next three where **none** of the planes are parallel or coincident. The very good news: you can determine the answer *just by looking* at the **row reduced** equations. It's easy! See page 47.

Planes E1, E2, E3 (with normals $\vec{n_1}, \vec{n_2}, \vec{n_3}$) ...	... which, row reduced*, become... *See page 47.	Geometrical description	Picture	What are multiples of what?
(6) $x+y+z=1$ $x+2y+2z=2$ $2x+3y+3z=4$	$x+y+z=1$ $y+z=1$ $0=1$	None are ∥. There are 3 ∥ lines of intersection!		none
(7) $x+y+z=1$ $2x+3y+3z=3$ $x+2y+2z=2$	$x+y+z=1$ $y+z=1$ $0=0$	None are ∥. There is one common line of intersection.		none
(8) $x+y+z=1$ $2x+3y+3z=4$ $x+2y+3z=2$	$x+y+z=1$ $y+z=2$ $z=-1$	None are ∥. There is a unique intersection point.		none

Here is the row reduction for (6):

| E1: $x+y+z=1$
 E2: $x+2y+2z=2$
 E3: $2x+3y+3z=4$ | $\begin{array}{l}\text{E4}=\text{E1}\\ \text{E5}=-\text{E1}+\text{E2}\\ \text{E6}=-2\text{E1}+\text{E3}\end{array}$ $\Leftrightarrow$ | E4: $x+y+z=1$
 E5: $y+z=1$
 E6: $y+z=2$ | $\begin{array}{l}\text{E7}=\text{E4}\\ \text{E8}=\text{E5}\\ \text{E9}=-\text{E5}+\text{E6}\end{array}$ $\Leftrightarrow$ | E7: $x+y+z=1$
 E8: $y+z=1$
 E9: $0=1$ |

Now you do the row reductions for (7) and (8)!

Summation Notation and Common SUM $\equiv \sum$ Formulas

Let m and n be natural numbers with $1 \leq m \leq n$.

$$\sum_{i=1}^{n} f(i) = f(1) + f(2) + f(3) + ... + f(n) \qquad \sum_{i=m}^{n} f(i) = f(m) + f(m+1) + ... + f(n)$$

$$\sum_{i=1}^{n} cf(i) = c\sum_{i=1}^{n} f(i) \qquad \sum_{i=1}^{n} (f(i) \pm g(i)) = \sum_{i=1}^{n} f(i) \pm \sum_{i=1}^{n} g(i)$$

$$\sum_{i=1}^{n} c = cn \qquad \sum_{i=1}^{n} i = \frac{n(n+1)}{2} \qquad \sum_{i=1}^{n} i^2 = \frac{n(n+1)(2n+1)}{6} \qquad \sum_{i=1}^{n} i^3 = \frac{n^2(n+1)^2}{4} = \left(\sum_{i=1}^{n} i\right)^2$$

Example 1) Evaluate each of (a) and (b) and expand (c) into THREE sums.

(a) $\sum_{i=1}^{11} 2i$ (b) $\sum_{i=4}^{8} i^3$ (c) $\sum_{i=1}^{n} (1+i)^2$

Solution (a) $\sum_{i=1}^{11} 2i = 2\sum_{i=1}^{11} i = 2\left(\frac{11 \cdot 12}{2}\right) = 132$

(b) $\sum_{i=4}^{8} i^3 \overset{\boxed{\sum_{i=1}^{8} i^3 - \sum_{i=1}^{3} i^3}}{=} \left(\frac{8 \cdot 9}{2}\right)^2 - \left(\frac{3 \cdot 4}{2}\right)^2 = 36^2 - 36 = 1260$

(c) $\sum_{i=1}^{n} (1+i)^2 = \sum_{i=1}^{n} (1+2i+i^2) = \sum_{i=1}^{n} 1 + 2\sum_{i=1}^{n} i + \sum_{i=1}^{n} i^2$

Example 2) Write the following in summation notation and evaluate the sum:

$$S = \frac{2}{n}\left(\frac{2}{n}\right)^2 + \frac{2}{n}\left(\frac{4}{n}\right)^2 + \frac{2}{n}\left(\frac{6}{n}\right)^2 + ... + \frac{2}{n}\left(\frac{2i}{n}\right)^2 + ... + \frac{2}{n}\left(\frac{2n}{n}\right)^2$$

Solution $S = \frac{2}{n}\left(\frac{2}{n}\right)^2 + \frac{2}{n}\left(\frac{4}{n}\right)^2 + \frac{2}{n}\left(\frac{6}{n}\right)^2 + ... + \frac{2}{n}\left(\frac{2i}{n}\right)^2 + ... + \frac{2}{n}\left(\frac{2n}{n}\right)^2$

$$= \sum_{i=1}^{n} \frac{2}{n}\left(\frac{2i}{n}\right)^2 \overset{\boxed{\text{Factor out the } \frac{2}{n} \text{ and the } \left(\frac{2}{n}\right)^2!}}{=} \frac{2}{n}\left(\frac{2}{n}\right)^2 \sum_{i=1}^{n} i^2 = \frac{8}{n^3} \frac{n(n+1)(2n+1)}{6} = \frac{4(n+1)(2n+1)}{3n^2}$$

Note : for evaluating integrals from definition (where a sum like this usually arises),

we would continue $\dfrac{4(n+1)(2n+1)}{3n^2} = \dfrac{4}{3}\left(\dfrac{n+1}{n}\right)\left(\dfrac{2n+1}{n}\right) = \dfrac{4}{3}\left(1+\dfrac{1}{n}\right)\left(2+\dfrac{1}{n}\right)$.

Arithmetic and Geometric Sequences and Series

The nth term of an **arithmetic** sequence with first term a_1 and common difference d is $a_n = a_1 + (n-1)d$. The sum is $S_n = \sum_{i=1}^{n} a_i = \dfrac{n}{2}(2a_1 + (n-1)d) \stackrel{\text{also}}{=} n\left(\dfrac{a_1 + a_n}{2}\right)$.

The nth term of a **geometric** sequence with first term a_1 and common ratio r is $a_n = a_1 r^{n-1}$. The sum is $S_n = \sum_{i=1}^{n} a_1 r^{i-1} \stackrel{\text{Use this when }r>1.}{=} a_1\left(\dfrac{r^n - 1}{r - 1}\right) \stackrel{\text{Use this when }r<1.}{=} a_1\left(\dfrac{1 - r^n}{1 - r}\right)$.

Here are the corresponding formulas for i starting at 0.	
Arithmetic: $\begin{array}{l} a_n = a_0 + nd \\ S_n = \dfrac{n+1}{2}(a_0 + a_n) \end{array}$	Geometric: $\begin{array}{l} a_n = a_0 r^n \\ S_n = a_0\left(\dfrac{1 - r^{n+1}}{1 - r}\right) \end{array}$

Example 1) Given an arithmetic series with $a_1 = 7$ and common difference $d = 2$, find a_6 and S_6.

Solution $a_6 \stackrel{a_1=7,\, n=6,\, d=2}{=} a_1 + (6-1)2 = 7 + 5(2) = 17$

$S_6 \stackrel{a_1=7,\, n=6,\, d=2}{=} \dfrac{6}{2}(2(7) + (6-1)2) = 3(14 + 10) = 72$ **OR** $S_6 \stackrel{a_1=7,\, a_6=17}{=} \dfrac{6}{2}(7 + 17) = 72$

Example 2) Given a geometric sequence with first term $a_1 = 3$ and common ratio $r = 2$, find a_6 and S_6.

Solution $a_6 = a_1 r^5 \stackrel{a_1=3,\, r=2}{=} 3(2^5) = 96$ $\quad S_6 = a_1\left(\dfrac{r^6 - 1}{r - 1}\right) \stackrel{a_1=3,\, r=2}{=} 3\left(\dfrac{2^6 - 1}{2 - 1}\right) = 189$

Example 3) An arithmetic sequence has $a_4 = 16$ and $a_{12} = 56$. Find a_1 and d.

Solution $a_{12} = a_1 + 11d = 56$ and $a_4 = a_1 + 3d = 16$. Subtracting these two equations gives $8d = 40$ so $d = 5$. Substituting: $a_4 = a_1 + 15 = 16$ and so $a_1 = 1$.

Example 4) A geometric sequence has $a_4 = 10\,000$ and $a_7 = 10$. Find a_1 and r.

Solution $a_4 = a_1 r^3 = 10000$ and $a_7 = a_1 r^6 = 10$.

$\dfrac{a_7}{a_4} = \dfrac{a_1 r^6}{a_1 r^3} = r^3 = \dfrac{10}{10\,000} = \dfrac{1}{1000}$ and so $r = \dfrac{1}{10}$. Substituting in a_4 gives

$a_1\left(\dfrac{1}{10}\right)^3 = 10\,000$ and so $a_1 = 10\,000\,000$.

Combinations and Permutations: Choosing and Arranging

For any natural numbers n and r, where $n \geq r$,

$$C(n, r) \overset{\text{Another Notation!}}{=} \binom{n}{r} \overset{\text{Another Notation!!}}{=} {}_nC_r = \frac{n!}{(n-r)!\,r!} = \binom{n}{n-r} \text{ calculates}$$

the number of **COMBINATIONS** of r objects you can make from n objects.

$$P(n, r) \overset{\text{Another Notation!}}{=} {}_nP_r = \frac{n!}{(n-r)!} \text{ calculates}$$

the number of **ARRANGEMENTS** of r objects you can make from n objects.

Remember, $0! = 1! = 1$ and $i! = i(i-1)(i-2)\ldots(3)(2)(1)$.

$C(n, 0) = C(n, n) = 1 \qquad P(n, 0) = 1 \qquad P(n, n) = n!$

Example 1) How many possible ways are there of choosing a president, vice-president, and treasurer from a group of 8 candidates for a student math club? **(Yes, there are student math clubs!)**

Solution We are looking for the number of **arrangements** of 3 people from 8. **Order IS important!** The required number is $P(8, 3) = \dfrac{8!}{(8-3)!} = \dfrac{8!}{5!} \overset{8\times 7 \times 6}{=} 336$.

Example 2) How many possible ways are there of choosing 3 members for the executive of the math club from 8 candidates?

Solution We are looking for the number of **combinations** of 3 people from 8. **Order is NOT important!**

The required number is $C(8, 3) = \dfrac{8!}{(8-3)!\,3!} = \dfrac{8!}{5!\,3!} \overset{8 \times 7}{=} 56$.

One (BIG TABLE) For You – Intersection of Three Planes: Non-Parallel/Non-Coincident

1) Complete the table.

Three planar equations…	… when row reduced, become…	Geometrical description	Picture
E1: $x - 2y - z = 1$ E2: $x - y - 2z = 2$ E3: $3x - 5y + z = -1$			
E1: $x - 2y - z = 1$ E2: $x - y - 2z = 2$ E3: $5x - 8y - 7z = 7$			
E1: $x + y + z = 1$ E2: $x + 2y + 2z = 2$ E3: $3x + 4y + 4z = 5$			

Answer

Three planar equations…	… when row reduced, become…	Geometrical description	Picture
E1: $x - 2y - z = 1$ E2: $x - y - 2z = 2$ E3: $3x - 5y + z = -1$	$x - 2y - z = 1$ $y - z = 1$ $z = -1$	None are ∥. There is a unique intersection point.	
E1: $x - 2y - z = 1$ E2: $x - y - 2z = 2$ E3: $5x - 8y - 7z = 7$	$x - 2y - z = 1$ $y - z = 1$ $0 = 0$	None are ∥. There is one common line of intersection.	
E1: $x + y + z = 1$ E2: $x + 2y + 2z = 2$ E3: $3x + 4y + 4z = 5$	$x + y + z = 1$ $y + z = 1$ $0 = 1$	None are ∥. There are 3 ∥ lines of intersection!	

Two For You – Summation Notation and Common Sum Formulas

1) Evaluate: $\sum_{i=1}^{10}(2i - i^2)$

2) Expand into THREE sums: $\sum_{i=1}^{n}\frac{3}{n}\left(1 + \frac{3i}{n}\right)^2$

Answers 1) -275 2) $\frac{3}{n}\sum_{i=1}^{n}1 + \frac{18}{n^2}\sum_{i=1}^{n}i + \frac{27}{n^3}\sum_{i=1}^{n}i^2$

Two For You – Arithmetic and Geometric Sequences and Series

1) Given an arithmetic sequence with 5th term 50 and 13th term 26, find a_1 and d.

2) Given a geometric sequence with 5th term 2 and 12th term 256, find a_1 and r.

Answers 1) $d = -3$ and $a_1 = 62$ 2) $r = 2$ and $a_1 = \dfrac{1}{8}$

Two For You – Combinations and Permutations

1) How many different poker hands (5 cards from a 52 card deck) contain the Ace and Jack of Spades?

2) For the word "utopia", how many possible ways are there of arranging

(a) exactly 4 of all the letters?   (b) just the vowels?

Answers 1) $C(50,3) = 19600$ 2)(a) $P(6,4) = 360$ (b) $P(4,4) = 24$

Elementary Probability

Once you read through this page, there is a good chance you will never buy another lottery ticket. The method to calculate the probability of an event is to take the number of ways an event can happen and divide it by the total number of possible outcomes. Calculating the number of outcomes can be tricky. We will keep things fairly basic here, but it would be a good idea to review combinations and permutations on page 252.

Example 1) Calculate the probability of drawing an ace in a standard 52 card deck.

Solution There are 4 aces in the deck, so the probability is $\frac{4}{52} = \frac{1}{13} \doteq 7.7\%$.

Example 2) A bag of marbles contains 3 red marbles and 2 green marbles. If you reach in and pull out two marbles, what is the probability of drawing both green marbles?

Solution Since the order of the two marbles does not matter, we will work with combinations here. There is one way for this event to happen: draw the two green marbles. There are $C(5,2) = \frac{5!}{(5-2)!2!} = 10$ possible outcomes. Therefore the probability is $\frac{1}{10} = 10\%$.

Example 3) A poker hand consists of 5 cards dealt from a standard 52 card deck. The highest hand is called a Royal Flush, where you have an Ace, King, Queen, Jack and Ten all from the same suit. What is the probability of being dealt a Royal Flush?

Solution There are four possible ways to get a Royal Flush (one for each suit). There are $C(52,5)$ possible poker hands.
$$C(52,5) = \frac{52!}{(52-5!)5!} = \frac{52 \times 51 \times 50 \times 49 \times 48}{5 \times 4 \times 3 \times 2 \times 1} = 2\,598\,960$$
Therefore, the probability is $\frac{4}{2\,598\,960} = \frac{1}{649\,740} \doteq 0.000154\%$.

Example 4) A lottery is played by choosing 6 numbers from 1 to 49. If you match all 6 winning numbers, you win the jackpot (order does not matter).

Solution There is only one way to win: you have all six numbers correct. The total number of possible outcomes is given by
$$C(49,6) = \frac{49!}{(49-6)!6!} = \frac{49 \times 48 \times 47 \times 46 \times 45 \times 44}{6 \times 5 \times 4 \times 3 \times 2 \times 1} = 13\,983\,816.$$ This means the probability of winning is $\frac{1}{13\,983\,816} \doteq 0.00000715\%$. Do you still want to buy that ticket?

Mean, Median, Mode, and Standard Deviation

Given numbers $X_1 \leq X_2 \leq X_3 \leq ... \leq X_n$, we have **THREE** kinds of averages:

$$\textbf{Mean} = \frac{X_1 + X_2 + X_3 + ... + X_n}{n} = \frac{\sum_{i=1}^{n} X_i}{n}$$

Median The numbers have been listed from lowest to highest. The median is

$$\begin{cases} \text{the middle number } X_{\frac{n+1}{2}}, \text{ if } n \text{ is odd;} \\ \text{the average } \dfrac{X_{\frac{n}{2}} + X_{\frac{n}{2}+1}}{2}, \text{ if } n \text{ is even.} \end{cases}$$

Mode The mode is the number that occurs most often in the list.
There may be several modes.

The **Standard Deviation** is a measure of how the data is scattered about the mean.
If the mean = $\bar{X}$, then

$$\textbf{Standard Deviation} = \sqrt{\frac{\sum_{i=1}^{n}(X_i - \bar{X})^2}{n}}. \text{ Note that } \frac{\sum_{i=1}^{n}(X_i - \bar{X})^2}{n} \text{ is itself a "mean".}$$

It is the **MEAN of the squares of the distances of the data points to the MEAN of the original data.**

Example 1) A group of 13 customers in a women's shoe store have these shoe sizes: 6, 6.5, 7, 7, 7, 7.5, 7.5, 8, 8, 8, 9, 9, 10. Find the mean, median, mode(s), and standard deviation for this data.

Solution $\textbf{Mean} = \dfrac{6 + 6.5 + 7 \times 3 + 7.5 \times 2 + 8 \times 3 + 9 \times 2 + 10}{13} = \bar{X} \doteq 7.7$

Median $= X_{\frac{13+1}{2}} = X_7 = 7.5$

Mode There are two modes: both 7 and 8 occur three times in the list.

The **Standard Deviation**

$$= \sqrt{\frac{(6-\bar{X})^2 + (6.5-\bar{X})^2 + 3(7-\bar{X})^2 + 2(7.5-\bar{X})^2 + 3(8-\bar{X})^2 + 2(9-\bar{X})^2 + (10-\bar{X})^2}{13}}$$

$\doteq 1.07$

The Binomial Theorem

For any natural number n,

$$(a+b)^n = \sum_{i=0}^{n} \binom{n}{i} a^{n-i} b^i$$

$$= \binom{n}{0} a^n + \binom{n}{1} a^{n-1} b + \binom{n}{2} a^{n-2} b^2 + \ldots + \binom{n}{i} a^{n-i} b^i + \ldots + \binom{n}{n-1} a^1 b^{n-1} + \binom{n}{n} b^n,$$

where $\binom{n}{i} = \dfrac{n!}{(n-i)!\,i!} = \binom{n}{n-i}$, and in particular, $\binom{n}{0} = \binom{n}{n} = 1$.

Remember, $0! = 1! = 1$ and $i! = i(i-1)(i-2)\ldots(3)(2)(1)$.

Some students think "3!" means **"THREE"**

Example 1) Expand $(2+x)^5$.

Solution $(2+x)^5 = \binom{5}{0} 2^5 + \binom{5}{1} 2^4 x + \binom{5}{2} 2^3 x^2 + \binom{5}{3} 2^2 x^3 + \binom{5}{4} 2 x^4 + \binom{5}{5} x^5$

$= 32 + 5(16)x + 10(8)x^2 + 10(4)x^3 + 5(2)x^4 + x^5$

$= 32 + 80x + 80x^2 + 40x^3 + 10x^4 + x^5$

Example 2) Find the coefficient of x^{15} in the expansion of $\left(2x^3 + \dfrac{1}{4x^2}\right)^{10}$.

Solution $\left(2x^3 + \dfrac{1}{4x^2}\right)^{10} = \sum_{i=0}^{10} \binom{10}{i} (2x^3)^{10-i} \left(\dfrac{1}{4x^2}\right)^i$

$\boxed{\text{Expand, using properties of exponents.}} = \sum_{i=0}^{10} \binom{10}{i} \dfrac{2^{10-i} x^{30-3i}}{4^i x^{2i}} \quad \boxed{\text{Rewrite } 4^i \text{ in base 2.}} = \sum_{i=0}^{10} \binom{10}{i} \dfrac{2^{10-i} x^{30-3i}}{2^{2i} x^{2i}}$

$\boxed{\text{Combine exponents!}} = \sum_{i=0}^{10} \binom{10}{i} 2^{10-3i} x^{30-5i}.$

We want $30 - 5i = 15$. $\therefore -5i = -15$ and so $i = 3$.

The required coefficient is

$\binom{10}{3} 2^{10-3(3)} = \dfrac{10!}{7!\,3!} 2^1 = \dfrac{(10)(9)(8)}{(3)(2)(1)}(2) = 240.$

Proof by Induction

Think about a row of dominoes. Suppose you know that the first domino falls over. What if you also knew that the dominos were close enough together so that if one domino falls, it will knock down the next. What conclusion could you draw about the row of dominos? Every domino will fall! This simple idea is exactly how proof by induction works.

Example 1) Prove $\sum_{i=1}^{n} i = 1+2+3+...+n$ evaluates to $\dfrac{n(n+1)}{2}$ for all whole numbers n.

Solution First we need to prove that it works for the first number. This is called the basis or base case. In the domino analogy, this is showing that the first domino falls.

For $n=1$, Left Side $= \sum_{i=1}^{n} i = \sum_{i=1}^{1} i = 1$ and Right Side $= \dfrac{n(n+1)}{2} = \dfrac{1(1+1)}{2} = 1$.

Therefore the formula does work for $n=1$.

Now we need to show that if it works for some number k it will work for the next number $k+1$. In the domino analogy, this is showing that if one domino falls, it knocks down the next one. This is called the **inductive** step.

Assume $\sum_{i=1}^{k} i = \dfrac{k(k+1)}{2}$ works for some number k.

We need to prove that $\sum_{i=1}^{k+1} i = \dfrac{(k+1)[(k+1)+1]}{2} = \dfrac{(k+1)(k+2)}{2}$.

Left Side $= \sum_{i=1}^{k+1} i = \sum_{i=1}^{k} i + (k+1)$

$\overset{\text{by our assumption}}{=} \dfrac{k(k+1)}{2} + (k+1)$

$\overset{\text{make a common denominator}}{=} \dfrac{k(k+1) + 2(k+1)}{2}$

$\overset{\text{Notice the common factor of } k+1.}{=} \dfrac{(k+1)(k+2)}{2}$

$= $ Right Side

Therefore, if the formula works for k it works for $k+1$. In other words, if one domino falls, it knocks down the next one. Therefore, by induction, we know this formula will work for any whole number! Every domino falls!

Two For You – Elementary Probability

1) Find the probability of the following events:
(a) rolling an even number on a standard six-sided die.
(b) being dealt a Royal Flush from a Euchre deck. (Note: a Euchre deck contains 24 cards: a 9, 10, Jack, Queen, King and Ace from each of the four suits.)
(c) reaching into a hat that contains each letter of the alphabet (once) and drawing out in order M, A, T, H. Hint: since order does matter here, you should use permutations, not combinations.

2) Come up with an event that has a probability of: (a) 1   (b) 0.

Answers 1)(a) $\frac{3}{6} = \frac{1}{2} = 50\%$ (b) $\frac{4}{42\,504} = \frac{1}{10\,626} \doteq 0.009\%$ (c) $\frac{1}{358\,800} \doteq 0.0003\%$

2)(a) Any event that **must** happen, such as, rolling a number smaller than 10 on a standard six-sided die.
(b) Any **impossible** event, such as, being dealt five aces from a standard deck of cards.

One For You – Mean, Median, Mode, and Standard Deviation

1) Find
(a) the mean   (b) the median   (c) the mode(s)   (d) the standard deviation
for the data points 1, 2, 3, 3, 3, 3, 4, 4, 5, 8.

Answers 1)(a) 3.6   (b) $\frac{X_5 + X_6}{2} = 3$ (c) 3   (d) 1.8

Two For You – The Binomial Theorem

1) Expand: $\left(\dfrac{2}{x^2} - \dfrac{y}{3}\right)^4$

2) Find the coefficient of x^{-4} in the expansion of $\left(\dfrac{3}{x^2} + \dfrac{x^2}{9}\right)^4$.

Answers 1) $\dfrac{16}{x^8} - \dfrac{32y}{3x^6} + \dfrac{8y^2}{3x^4} - \dfrac{8y^3}{27x^2} + \dfrac{y^4}{81}$ 2) 12

Two For You – Proof by Induction

Prove the following using induction:

1) $\displaystyle\sum_{i=1}^{n} \dfrac{1}{i(i+1)} = \dfrac{n}{n+1}$

2) $\displaystyle\sum_{i=0}^{n} 2^i = 2^{n+1} - 1$

Answers 1) For $n=1$, Left Side = Right Side = $\dfrac{1}{2}$. Assume for $n=k$, that $\displaystyle\sum_{i=1}^{k}\dfrac{1}{i(i+1)} = \dfrac{k}{k+1}$.

Now prove $\displaystyle\sum_{i=1}^{k+1}\dfrac{1}{i(i+1)} = \dfrac{k+1}{k+2}$.

2) For $n=0$, Left Side = Right Side = 1. Assume for $n=k$, that $\displaystyle\sum_{i=0}^{k} 2^i = 2^{k+1} - 1$.

Now prove $\displaystyle\sum_{i=0}^{k+1} 2^i = 2^{k+2} - 1$.

INDEX

Topic	Page
absolute value	65, 66, 69, 70
absolute value equations	66
absolute value inequalities	69, 70
adding and subtracting fractions	4
$a^n \pm b^n$ (factoring)	15
angles (basics)	112
angles in standard position	120, 123
area using integrals	210
arctrigonometric graphs	141
arithmetic sequences and series	251
asymptotes (horizontal, vertical and slant/oblique)	87, 88
BEDMAS (order of operations)	3
binomial theorem	257
branch functions (continuity & differentiability)	162, 168
chain rule	184
chain rule in reverse	192, 195-198
combinations and permutations	252, 255
common factors	16
completing the square	58
complex numbers	10
composite functions	159
continuity and branch functions	162
continuity at a point	160
continuous functions	161
$\cos x = c$ (solving)	135
cosine (circle definition)	131
cosine law	137
critical numbers of a functions	171
cross or vector product of two vectors	237
decimals	6
definite integrals	210, 213
dependent linear systems	51
derivative from the definition ("first principles")	166
derivative of an integral	216
derivative of exponential functions	105
derivative of logarithm functions	106
derivatives (using the formulas)	105-108, 183-186, 189
derivatives from parametric equations	225, 226
derivatives of inverse functions	221
difference of cubes (factoring)	12
difference of squares (factoring)	11
differentiability and branch functions	168

differentiable functions	167
differential equations	219
differential (estimating functions values)	178
discontinuity at a point	160
distance between two points	36
distance from a point to a line or plane	36
dividing and multiplying fractions	5
domain of a function	156, 159
dot or scalar or inner product of two vectors	233, 234
easy (simple) trinomial factoring	17
elimination method for solving linear equations	45, 46
equation of a line	33, 238
equation of a plane	239, 240
essential discontinuities of a function	165
estimating using the differential	178
exponent properties	94
exponential derivatives	105
exponential equations	101
exponential graphs	96
extreme points using the first derivative	172
factor theorem	18
factoring	11, 12, 15-17, 53
factoring with the quadratic formula	53
"first principles" (derivative from the definition)	166
floor or greatest integer function	90
fraction arithmetic	4, 5
geometric sequences and series	251
graphing polynomials without calculus	84
graphing using calculus	174, 177
graphs	35, 76-78, 81-84, 93, 96, 99, 130, 141, 173, 174, 177
graphs (rescaling or shifting)	82
greatest integer function (graphs)	93
greatest integer or floor function	90
horizontal asymptotes	87
implicit differentiation	186, 189
improper integrals	214, 215
inconsistent linear systems	51
induction	258
inequalities	59, 60, 63, 64, 69, 70
inequalities with two or more factors	63
inner or dot or scalar product of two vectors	233, 234
integrals	111, 190-192, 195-198, 201-204, 207-210, 213-216
integration by partial fractions	209
integration by parts	201-203
integration by trigonometric substitution	204, 207, 208

integration of trig products	198
intersection of lines	45, 46, 243, 244
intersection of planes	47, 48, 51, 245, 246, 249
intersection of two curves	89
inverse formulas for exponents & logarithms	100
inverse trigonometric graphs	141
inverse of a function	220
L'Hôpital's Rule	154, 155
limits	142-144, 147-150, 153-155
linear inequalities	59
linear graphs	35
linear systems in two and three unknowns	45-48, 51
lines (parallel and perpendicular)	40, 112, 243
logarithm equations	102
logarithm graphs	99
logarithm properties	95
logarithmic differentiation	107, 108
maximum points from the first derivative	172
mean value theorem	180
median	256
minimum points from the first derivative	172
mode	256
multiplying and dividing fractions	5
normal line to a curve	41
oblique/slant asymptotes	88
order of operations (BEDMAS)	3
parabolas (graph of $y = a(x-b)^2 + c$)	57
parallel lines	40, 112, 243
parametric equations	222, 225, 226
partial fractions	27-30, 209
permutations & combinations	252, 255
percent	9
perpendicular lines	40
planes, equations of	239, 240
planes, intersections of	245, 246, 249
polar coordinates	227, 228, 231
polar coordinates to rectangular coordinates	228
polar equations to rectangular equations	231
polynomial division	24
polynomial fractions (adding and subtracting)	22
polynomial fractions (multiplying and dividing)	23
polynomial graphs without calculus	84
polynomial multiplication	21
probability	255

product rule	183
products and sum of roots of a quadratic equation	54
Pythagorean theorem	113
quadratic equations	52
quadratic formula	10, 52-54
quadratic graphs	76
quadratic inequalities	60
quotient rule	185
radian measure of an angle	117
rate	9
ratio	9
rational inequalities	64
rationalizing denominators that have $\sqrt{}$	75
rectangular coordinates to polar coordinates	231
rectangular equations to polar equations	228
related angles in standard position	123
remainder theorem	18
removable discontinuities	165
Rolle's theorem	179
row reduction method for solving linear equations	46-48
scalar equation of a plane	240
scalar or dot or inner product of two vectors	233, 234
separation of variables	219
similar triangles	114
simple (easy) trinomial factoring	17
$\sin x = c$	132
sine (circle definition)	131
sine law	136
slant/oblique asymptotes	88
slope and y intercept	34
slopes of lines	39
slopes (visually identifying)	39
SOH CAH TOA	118, 119
solving linear equations	42, 45-48, 51
square root	71, 72, 75
square root equations	72
standard deviation	256
substitution method for solving linear equations	45
subtracting and adding fractions	4
sum and products of roots of a quadratic equation	54
summation notation and formulas	250
symmetry tests	83
tangent line to a function	41
trigonometric equation $\cos x = c$ (solving)	135

trigonometric equation $\sin x = c$ (solving)	132
trigonometric formulas	138
trigonometric graphs	130
trigonometric ratios	118, 124-126, 129
trigonometric ratios for important angles	124-126, 129
trigonometric ratios for the (30°, 60°, 90°) triangle	124
trigonometric ratios for the (45°, 45°, 90°) triangle	125
trinomials (factoring)	17, 53
vector equation of a line	238
vector equation of a plane	239
vector or cross product of two vectors	237
vector projection	234
vectors (basics)	232
vertical asymptotes	87
y intercept and slope	34
y versus y' versus y''	173
$y = x^{-n}$ graphs	78
$y = x^n$ graphs	77
$y = x^{1/n}$ graphs	81

Almost Every Integration Formula You'll Ever Need!

Fill in each formula on the right using the corresponding formula on the left.

BASIC FORMULA	CHAIN RULE IN REVERSE
$\int k\, dx = kx + C$	
$\int k f'(x)\, dx = k f(x) + C$	$u' = du/dx$
$\int f(x) \pm g(x)\, dx = \int f(x)\, dx \pm \int g(x)\, dx$	↓↓↓↓↓
If $\int f'(x)\, dx = f(x) + C$ then...............	$\int f'(u) u'\, dx = f(u) + C$
$\int \sin(x)\, dx = -\cos(x) + C$	$\int \sin(u) u'\, dx = -\cos(u) + C$
$\int \cos(x)\, dx = \sin(x) + C$	$\int \cos(u) u'\, dx =$
$\int \sec^2(x)\, dx = \tan(x) + C$	$\int \sec^2(u) u'\, dx =$
$\int \csc(x) \cot(x)\, dx = -\csc(x) + C$	$\int \csc(u) \cot(u) u'\, dx =$
$\int \sec(x) \tan(x)\, dx = \sec(x) + C$	$\int \sec(u) \tan(u) u'\, dx =$
$\int \csc^2(x)\, dx = -\cot(x) + C$	$\int \csc^2(u) u'\, dx =$
$\int x^n\, dx = x^{n+1}/(n+1) + C$, $n \neq -1$	$\int u^n u'\, dx =$, $n \neq -1$
$\int e^x\, dx = e^x + C$	$\int e^u u'\, dx =$
$\int 1/x\, dx = \ln|x| + C$	$\int u'/u\, dx =$
$\int a^x\, dx = a^x/\ln a + C$	$\int a^u u'\, dx =$
$\int \tan(x)\, dx = -\ln|\cos(x)| + C$	$\int \tan(u) u'\, dx =$
$\int \cot(x)\, dx = \ln|\sin(x)| + C$	$\int \cot(u) u'\, dx =$
$\int \sec(x)\, dx = \ln|\sec(x) + \tan(x)| + C$	$\int \sec(u) u'\, dx =$
$\int \csc(x)\, dx = \ln|\csc(x) - \cot(x)| + C$	$\int \csc(u) u'\, dx =$

BASIC FORMULA	CHAIN RULE IN REVERSE
$\int \dfrac{1}{\sqrt{1-x^2}}\,dx = \arcsin(x) + C$	$\int \dfrac{u'}{\sqrt{1-u^2}}\,dx =$
$\int \dfrac{1}{1+x^2}\,dx = \arctan(x) + C$	$\int \dfrac{u'}{1+u^2}\,dx =$
$\int \dfrac{1}{|x|\sqrt{x^2-1}}\,dx = \operatorname{arcsec}(x) + C$	$\int \dfrac{u'}{|u|\sqrt{u^2-1}}\,dx =$
$\int \dfrac{1}{\sqrt{1+x^2}}\,dx = \operatorname{arcsinh}(x) + C$	$\int \dfrac{u'}{\sqrt{1+u^2}}\,dx =$
$\int \dfrac{1}{1-x^2}\,dx = \operatorname{arctanh}(x) + C$	$\int \dfrac{u'}{1-u^2}\,dx =$
$\int \tanh(x)\,dx = \ln(\cosh(x)) + C$ (Note no absolute value!)	$\int \tanh(u)\,u'\,dx =$
$\int \coth(x)\,dx = \ln|\sinh(x)| + C$	$\int \coth(u)\,u'\,dx =$

Every Exact Trig Ratio You'd Ever Want to Know and Probably More!

degrees	radians	sin	cos	tan	csc	sec	cot
0	0	0	1	0	undefined	1	undefined
30	$\frac{\pi}{6}$	$\frac{1}{2}$	$\frac{\sqrt{3}}{2}$	$\frac{1}{\sqrt{3}}$	2	$\frac{2}{\sqrt{3}}$	$\sqrt{3}$
45	$\frac{\pi}{4}$	$\frac{1}{\sqrt{2}}$	$\frac{1}{\sqrt{2}}$	1	$\sqrt{2}$	$\sqrt{2}$	1
60	$\frac{\pi}{3}$	$\frac{\sqrt{3}}{2}$	$\frac{1}{2}$	$\sqrt{3}$	$\frac{2}{\sqrt{3}}$	2	$\frac{1}{\sqrt{3}}$
90	$\frac{\pi}{2}$	1	0	undefined	1	undefined	0
120	$\frac{2\pi}{3}$	$\frac{\sqrt{3}}{2}$	$-\frac{1}{2}$	$-\sqrt{3}$	$\frac{2}{\sqrt{3}}$	-2	$-\frac{1}{\sqrt{3}}$
135	$\frac{3\pi}{4}$	$\frac{1}{\sqrt{2}}$	$-\frac{1}{\sqrt{2}}$	-1	$\sqrt{2}$	$-\sqrt{2}$	-1
150	$\frac{5\pi}{6}$	$\frac{1}{2}$	$-\frac{\sqrt{3}}{2}$	$-\frac{1}{\sqrt{3}}$	2	$-\frac{2}{\sqrt{3}}$	$-\sqrt{3}$
180	π	0	-1	0	undefined	-1	undefined
210	$\frac{7\pi}{6}$	$-\frac{1}{2}$	$-\frac{\sqrt{3}}{2}$	$\frac{1}{\sqrt{3}}$	-2	$-\frac{2}{\sqrt{3}}$	$\sqrt{3}$
225	$\frac{5\pi}{4}$	$-\frac{1}{\sqrt{2}}$	$-\frac{1}{\sqrt{2}}$	1	$-\sqrt{2}$	$-\sqrt{2}$	1
240	$\frac{4\pi}{3}$	$-\frac{\sqrt{3}}{2}$	$-\frac{1}{2}$	$\sqrt{3}$	$-\frac{2}{\sqrt{3}}$	-2	$\frac{1}{\sqrt{3}}$
270	$\frac{3\pi}{2}$	-1	0	undefined	-1	undefined	0
300	$\frac{5\pi}{3}$	$-\frac{\sqrt{3}}{2}$	$\frac{1}{2}$	$-\sqrt{3}$	$-\frac{2}{\sqrt{3}}$	2	$-\frac{1}{\sqrt{3}}$
315	$\frac{7\pi}{4}$	$-\frac{1}{\sqrt{2}}$	$\frac{1}{\sqrt{2}}$	-1	$-\sqrt{2}$	$\sqrt{2}$	-1
330	$\frac{11\pi}{6}$	$-\frac{1}{2}$	$\frac{\sqrt{3}}{2}$	$-\frac{1}{\sqrt{3}}$	-2	$\frac{2}{\sqrt{3}}$	$-\sqrt{3}$
360	2π	0	1	0	undefined	1	undefined

$\sqrt{2} \doteq 1.4142 \quad \frac{1}{\sqrt{2}} \doteq 0.7071 \quad \sqrt{3} \doteq 1.7321 \quad \frac{1}{\sqrt{3}} \doteq 0.5774 \quad \frac{\sqrt{3}}{2} \doteq 0.8660 \quad \frac{2}{\sqrt{3}} \doteq 1.1547$

The Rationale Behind Fractions

$$\frac{a}{b} + \frac{c}{d} = \frac{ad+bc}{bd} \qquad \frac{\left(\frac{a}{b}\right)}{\left(\frac{c}{d}\right)} = \frac{a}{b} \times \frac{d}{c} \qquad \frac{\left(\frac{a}{b}\right)}{c} = \frac{a}{bc} \qquad \frac{a}{\left(\frac{b}{c}\right)} = \frac{ac}{b}$$

Factoring Productively

$$ax + ay = a(x+y) \qquad x^2 - y^2 = (x-y)(x+y) \qquad x^3 \pm y^3 = (x \pm y)(x^2 \mp xy + y^2)$$

$$x^n - y^n = (x-y)(x^{n-1} + x^{n-2}y + x^{n-3}y^2 + \ldots + xy^{n-2} + y^{n-1}), \text{ for } n \in \mathbb{N}$$

$$x^n + y^n = (x+y)(x^{n-1} - x^{n-2}y + x^{n-3}y^2 - \ldots - xy^{n-2} + y^{n-1}), \text{ for } n \in \mathbb{N}, n \text{ odd}$$

The Power of Binomials

$$(x \pm y)^2 = x^2 \pm 2xy + y^2 \qquad (x \pm y)^3 = x^3 \pm 3x^2y + 3xy^2 \pm y^3$$

Advocating Exponents

$$a^x a^y = a^{x+y} \qquad \frac{a^x}{a^y} = a^{x-y} \qquad (a^x)^y = a^{xy} \qquad \left(\frac{ab}{c}\right)^x = \frac{a^x b^x}{c^x} \qquad a^0 = 1 \qquad a^{-1} = \frac{1}{a}$$

Getting Powerful with Logs

$$\log_a(xy) = \log_a(x) + \log_a(y) \qquad \log_a\left(\frac{x}{y}\right) = \log_a(x) - \log_a(y) \qquad \log_a(x^y) = y\log_a(x)$$

$$\log_a(x^y) \overset{\text{Please note!}}{\neq} [\log_a(x)]^y \qquad \log_a(1) = 0 \qquad \log_a(a) = 1 \qquad \log_a\left(\frac{1}{a}\right) = -1$$

Inverse formulas: $a^{\log_a(x)} = x \qquad \log_a(a^x) = x$ **Change of base:** $\log_a(x) = \frac{\log_b(x)}{\log_b(a)} \qquad \log_a(b) = \frac{1}{\log_b(a)}$

Rooting for the Quadratic Formula

$$ax^2 + bx + c = 0 \Rightarrow x = \frac{-b \pm \sqrt{b^2 - 4ac}}{2a} \qquad \text{Sum of roots} = -\frac{b}{a} \qquad \text{Product of roots} = \frac{c}{a}$$

The Slant on Slopes and Lines

Slope through points (x_1, y_1) and $(x_2, y_2) = \frac{y_2 - y_1}{x_2 - x_1}$; line $l_1 \perp$ line $l_2 \Rightarrow$ slope of $l_1 = -\frac{1}{\text{slope of } l_2}$.

The equation of the line through (x_1, y_1) with slope m: $y - y_1 = m(x - x_1)$.

Trig Truths

π radians $= 180°$ $\qquad \sin^2(A) + \cos^2(A) = 1 \qquad 1 + \tan^2(A) = \sec^2(A) \qquad \cot^2(A) + 1 = \csc^2(A)$

$$\sin\left(\frac{\pi}{2} - A\right) = \cos(A) \qquad \cos\left(\frac{\pi}{2} - A\right) = \sin(A) \qquad \tan\left(\frac{\pi}{2} - A\right) = \cot(A)$$

$$\sin(-A) = -\sin(A) \qquad \cos(-A) = \cos(A) \qquad \tan(-A) = -\tan(A)$$

$$\sin(A \pm B) = \sin(A)\cos(B) \pm \cos(A)\sin(B) \qquad \sin(2A) = 2\sin(A)\cos(A) \qquad \sin^2(A) = \frac{1 - \cos(2A)}{2}$$

$$\cos(A \pm B) = \cos(A)\cos(B) \mp \sin(A)\sin(B) \qquad \cos(2A) = \cos^2(A) - \sin^2(A) \qquad \cos^2(A) = \frac{1 + \cos(2A)}{2}$$

Sine Law: $\frac{\sin(A)}{a} = \frac{\sin(B)}{b} = \frac{\sin(C)}{c}$

Cosine Law (eg): $a^2 = b^2 + c^2 - 2bc\cos(A)$

NOTES

NOTES

NOTES